ALLE ZEIT WACH 1842

40. Colloquium der Gesellschaft für Biologische Chemie
6.-8. April 1989 in Mosbach/Baden

Molecular Mechanisms of Hormone Action

Edited by
U. Gehring, E. Helmreich and G. Schultz

With 77 Figures

Springer-Verlag Berlin Heidelberg New York
London Paris Tokyo Hong Kong

Prof. Dr. ULRICH GEHRING
Institut für Biologische Chemie
Universität Heidelberg
Im Neuenheimer Feld 510
6900 Heidelberg, FRG

Prof. Dr. ERNST J.M. HELMREICH
Institut für Physiologische Chemie
Universität Würzburg
Koellikerstr. 2
8700 Würzburg, FRG

Prof. Dr. GÜNTER SCHULTZ
Institut für Pharmakologie
Freie Universität Berlin
Thielallee 67–73
1000 Berlin 33, FRG

ISBN-13: 978-3-642-75024-3 e-ISBN-13: 978-3-642-75022-9
DOI: 10.1007/978-3-642-75022-9

Library of Congress Cataloging-in-Publication Data.
Gesellschaft für Biologische Chemie. Colloqium (40th: 1989: Mosbach, Baden-Württemberg, Germany). Molecular mechanisms of hormone action / 40. Colloquium der Gesellschaft für Biologische Chemie, 06.–08. April 1989 in Mosbach/Baden; edited by U. Gehring, E. Helmreich, and G. Schultz. p. cm. 1. Molecular endocrinology-Congresses. 2. Hormones-Physiological effect-Congresses. 3. Hormone receptors-Congresses. I. Gehring, Ulrich. II. Helmreich, E. J. M. (Ernst J. M.), 1982– . III. Schultz, G. (Günter), 1936– . IV. Title. [DNLM: 1. Hormones-pharmacology-congresses. WK 102 G389m] QP187.3.M64G47 1989 615'.36-dc20 DNLM/DLC 89-21860

Softcover reprint of the hardcover 1st edition 1989

Typesetting: Overseas Typographers, Inc., Makati, Philippines

2131/3145-543210

Preface

The 40th Colloquium of the Gesellschaft für Biologische Chemie in Mosbach calls to mind the evolution of these meetings over the years which faithfully copies the explosive growth of biochemistry. Among the preceding 39 colloquia, which were held since 1949, there were only two devoted to hormone action. The first one was held in Mosbach from 30 April to 1 May 1954 under the auspices of the Gesellschaft für Physiologische Chemie, as the Society was named at that time. This colloquium dealt with *Hormone und ihre Wirkungsweise* (Hormones and Their Actions) and it was arranged by the Professor of Biochemistry at Frankfurt University, the late Kurt Felix, who at that time was also President of the Society. Major emphasis was on steroid hormones and the peptide hormones secreted by the anterior pituitary, the thyroid and the adrenal gland, but a paper, which received much attention, was that of Christian de Duve on *Le mode d'action de l'insuline* in which a controversy was discussed, i. e. an action of insulin at the enzyme level versus an action on the permeability of muscle for glucose. It was this lecture which influenced one of us (E.H.) to go in the same year to Carl F. Cori in St. Louis to work on this problem. A second colloquium on *Wirkungsmechanismen der Hormone* (Mode of Action of Hormones) was held in Mosbach 13 years later, exactly at the same time as this year's meeting. It was arranged by Peter Karlson. At this colloquium steroid hormone receptors, possible interactions with the nucleus and Gordon M. Tomkins' studies on enzyme induction in cells cultured in vitro were already introduced. Retrospectively, however, the highlight was a lecture by Earl W. Sutherland who discussed the role of cyclic 3′,5′-AMP in hormone action. It was at that occasion that G.S. met Earl W. Sutherland and subsequently joined him and Joel Hardman for 2 years at Vanderbilt University.

U.G. had originally proposed to the members of the Gesellschaft für Biologische Chemie to hold another meeting on hormone action, a proposal which was finally approved, and the 40th Colloquium was held in Mosbach on "Molecular Mechanisms of Hormone Action" from 6–8 April 1989.

The Proceedings of this Colloquium attest to the exponential growth of molecular endocrinology in the intervening 22 years. Among the spectacular advances is the progress in gene regulation and transcriptional control by steroid hormones. Research on steroid hormones has an impressive tradition in Germany established by Adolf Butenandt and

Peter Karlson. It was therefore appropriate that this year's meeting started with a historical review by Peter Karlson. Subsequently, P. Chambon, M. Beato, R. Renkawitz, B. Groner, S.M. Hollenberg, U. Gehring, G. Schütz und G. Hager discussed the interactions of steroid hormone receptors with chromatin, with DNA and with transcription factors. Moreover, among the topics under discussion were structure and function of steroid hormone receptors, the role of cAMP and steroid hormones on gene regulation, relationships between steroid hormone and thyroid hormone receptors, modulation of glucocorticoid action by oncogenes and peptide hormones and hormonal control of cell differentiation.

As impressive as the progress made in the field of steroid hormone action is the development initiated by the discoveries of cAMP and cGMP in Earl Sutherland's laboratory. This led to the characterization of cAMP-dependent protein kinase by E.G. Krebs and, following that, led to the discovery of other protein kinases, such as tyrosine kinase, casein kinase and protein kinase C, and of the corresponding phosphatases and their participation in complex-regulated phosphorylation-dephosphorylation cascades. The present state of the regulatory role of protein phosphorylation was discussed in an evening lecture by E.G. Krebs himself.

Another spin-off of the discovery of cyclic nucleotides as second messenger was the study of G-protein-coupled receptor and effector systems including visual signal transmission in rod outer segment membranes. In 1971 Martin Rodbell with Lutz Birnbaumer and co-workers had observed that GTP enhances glucagon stimulation of rat liver adenylate cyclase, although GTP is not a substrate of adenylate cyclase. Subsequently, the Würzburg group suggested that a separate guanine nucleotide binding protein might be the putative transducer postulated by M. Rodbell, an expectation which was realized in 1977 when T. Pfeuffer isolated a GTP binding protein which could be resolved from adenylate cyclase in a reversible fashion. At about the same time, D. Cassel and Z. Selinger ascribed a β-adrenergic and cholera toxin-regulated GTPase activity to the stimulating G_s-protein in turkey erythrocyte membranes. In the following 3 years, the GTP binding protein involved in hormonal activation of adenylate cyclase and that participating in visual signal transmission was purified to homogeneity and its subunit structure was clarified in the laboratories of A.G. Gilman, E. Ross and L. Birnbaumer. Work with the G-protein involved in visual signal transduction was carried out by L. Stryer with B.K.-K. Fung and in other laboratories. Thanks to a powerful combination of recombinant DNA technology and classical protein biochemistry, many more GTP binding proteins were discovered and cloned in the meantime, and their number is still growing. New functions were found and a role of GTP binding proteins in the control of other enzymes and targets aside from adenylate cyclase and cGMP phosphodiesterase became apparent. The list of speakers: Y. Kaziro, M.G. Caron, K.H. Jakobs, E. Ross, T. Pfeuffer, M. Chabre, who were invited to participate in this session on G-proteins, included also Dr. Irving S. Sigal. He had accepted our invitation and was supposed to report on his work on the ras oncogene protein. However, Dr. Sigal became a victim of the disaster of Lockerbee in Scotland. His absence was deeply felt by all of us. We, like the whole scientific community, mourn him. The Merck Sharp and Dohme Company enabled Dr. M. Marshall to report on the recent work from Dr. I.S. Sigal's laboratory.

The last years have seen the rapid establishment of a new second messenger system which was triggered by the discovery of M.J. Berridge and R.F. Irvine of inositol trisphosphate, the second messenger regulating intracellular Ca^{2+}. A session was devoted to phosphoinositide metabolism. We learned from Zvi Selinger about its role in visual signal transduction in *Drosophila* and were informed by S.C. Rhee on phospholipase C isoenzymes and by P. Parker on protein kinase C. John H. Exton found that phosphatidylcholine, in addition to phosphatidylinositol-phosphates, plays a major role in the hormonally regulated phospholipase-catalyzed phospholipid breakdown. H. Reuter, F. Hofmann and W. Rosenthal were the speakers of a session dealing with hormonally modulated ion channels, and Lutz Birnbaumer reported on the controversial involvement of α- and $\beta\gamma$-subunits of G-proteins in the hormonal regulation of potassium and calcium channels, an issue which has attracted much attention. The previously quite elusive role of cGMP, which like cAMP was discovered originally in Earl W. Sutherland's laboratory, is becoming clearer, mainly through its involvement in the action of the atrial natriuretic peptide. In the final session of the meeting D.L. Garbers and F. Murad informed us on guanylate cyclases, the enzymes which form cyclic GMP from GTP, an activity which was first described in 1969. Axel Ullrich's lecture on insulin and growth factor receptor-mediated signaling allowed the appreciation of how much can be achieved by recombinant DNA technology including site-directed mutagenesis and the use of chimeric hybrid receptors.

In summing up the 40th Mosbach colloquium one does not need to emphasize that the study of molecular mechanisms of hormone action has become one of the fastest growing fields in molecular and cellular biology with many implications for neurobiology, oncology, endocrinology, clinical research and industrial applications. Although this is a fact, it is tempting to speculate on what might be in store in the future. It is probably quite realistic to expect that in the field of steroid hormone action attention will be focused on gene regulation. Encouraged by the progress made recently in the study of repressor-DNA interactions, the next goal might well be the elucidation of the 3D-structure of a DNA-steroid hormone receptor complex. The equally attractive project, namely expression and crystallization of a membrane-bound G-coupled receptor, might be more difficult to tackle. The question is how to construct, by genetic manipulation, a superexpression system providing the milligram quantities of membranous receptor proteins required for X-ray structural work.

The success of a meeting always depends on the quality of the speakers. The fact that this meeting attracted an unexpectedly large number of interested participants is due to them and to nothing else. This we appreciate and are grateful to our speakers and their cooperation. For the first time a delegation of the „Gesellschaft für Biochemie der DDR" led by its President, Professor Harald Aurich from Halle, accompanied by Professor Krause from Berlin took part in a meeting of our Society. Let us hope that this signals a new beginning in the relationship between the two German biochemical societies.

We, the organizers, wish to thank the Gesellschaft für Biologische Chemie for support in organizing this meeting. Special thanks are due to the treasurer of the Society, Dr. E. Truscheit, who greatly helped to finance this meeting and to the secretaries, Mrs. E. Haring and Mrs. E.M. Preiss. The financial support of the Martini Foundation and all its sponsors and especially the support of the Deutsche Forschungsgemeinschaft, the VW Foundation and the Fonds der Chemischen Industrie e.V. is gratefully acknowledged. We also wish to thank the projectionists and the many other voluntary helpers, including the local authorities, who took care of the many technical details and carried the main burden of running a meeting of this size.

July 1989

Ernst Helmreich, Würzburg
Ulrich Gehring, Heidelberg
Günter Schultz, Berlin

Contents

Contributors

You will find the addresses at the beginning of the respective contribution

Part 1

Mode of Action of Steroid Hormones: A Historical Review

P. KARLSON[1]

1 Introduction

This autumn we can celebrate the 60th anniversary of the isolation of the first steroid hormone, estrone, by E.A. Doisy in the USA and – independently – A. Butenandt in Germany. Within 5 years the other sex hormones (androsterone, testosterone, progesterone) were isolated, followed by the glucocorticosteroids in 1936–39.

When the structure of these hormones was elucidated, biochemists found it exciting that the hormones were steroids, i.e., their structures were related to cholesterol. The latter was regarded, at that time, as a rather dull substance, being responsible for gallstones and present in cell membranes. Now it appeared that cholesterol was also the mother substance of such delicate things as steroid hormones. Butenandt (1936) gave a tentative scheme of the relationship between sex hormones and cholesterol (Fig. 1).

Though there was some interest in the biogenesis and also in the inactivation of steroid hormones at that time, the question of the mechanism of steroid hormone action is not mentioned in any of the contemporary reviews. And, indeed, for the isolation of a hormone only a reliable bioassay is needed; how this physiological effect is brought about is irrelevant.

2 Are Steroid Hormones Coenzymes or Effectors of Enzyme Activity?

Like hormones, vitamins are substances that are active in bioassays in very low amounts. In the late 1930s it was found that vitamins are part of coenzymes which in higher organisms cannot be biosynthesized in the body. It could be visualized that hormones act in a similar way and, indeed, in an important review, D.E. Green (1941) wrote:

"The thesis which we shall develop in this article is that any substance which occurs in traces in the cell and which is necessary in traces in the diet or medium must either be an essential part of some enzyme or an enzyme itself...
We may therefore in accordance with the trace substance -enzyme thesis consider hormones either as potential enzymes or as prosthetic groups."

[1]Institut für Molekularbiologie und Tumorforschung der Philipps-Universität, Emil-Mannkopff-Straße 2, D-3550 Marburg, Fed. Rep. of Germany

40. Colloquium Mosbach 1989
Molecular Mechanisms of Hormone Action

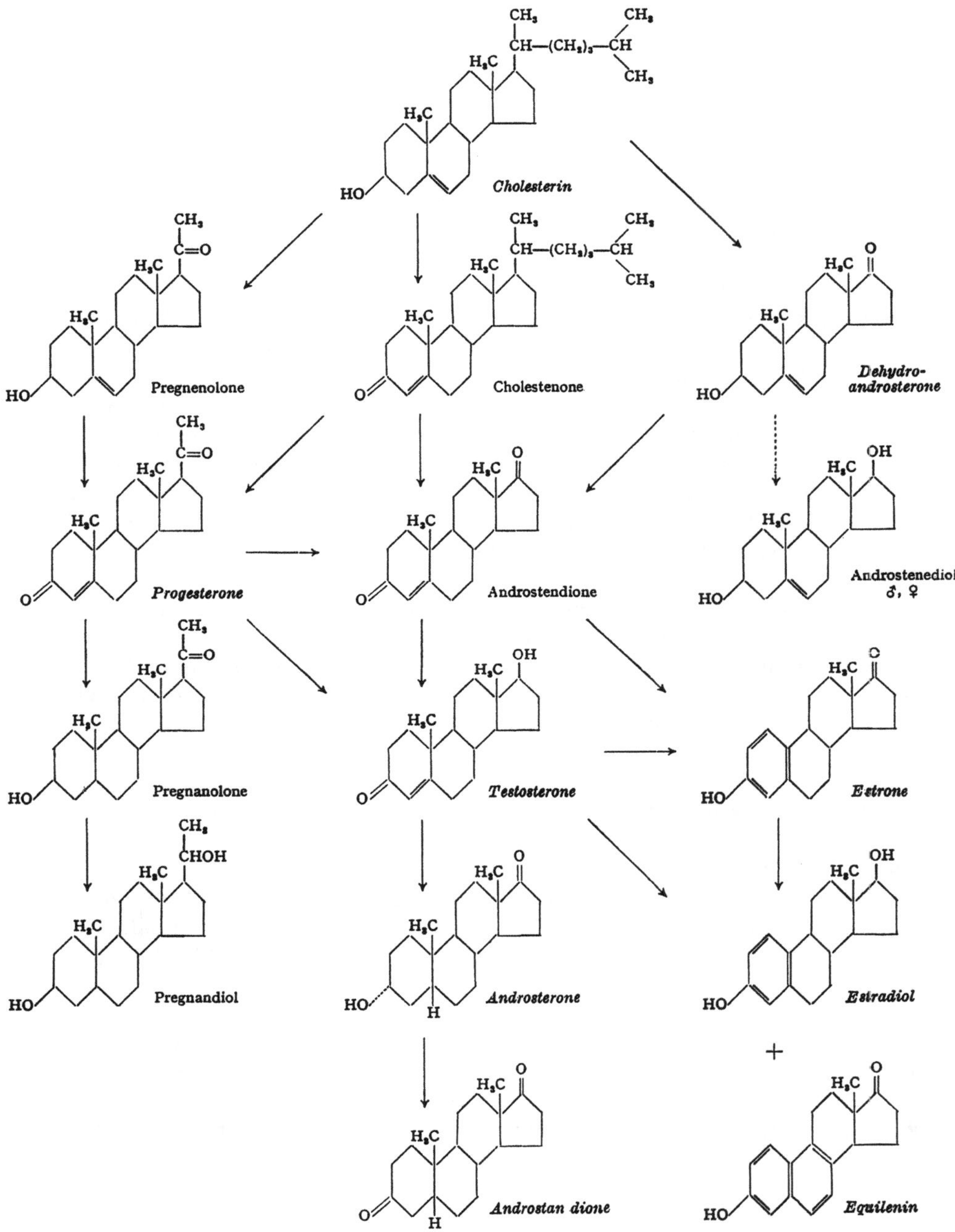

Fig. 1. Possible biogenetic relationship of the sex hormones. (Butenandt 1936)

For about 15 years, this remained the guiding hypothesis for the study of enzyme action. Many enzyme systems were subjected to the action of various steroid hormones either in vivo or in vitro. Small effects were sometimes found, either a stimulation or – more often – an inhibition, but nothing that could explain the physiological actions of these hormones which were well known.

In the late 1950s, the idea of a coenzyme-like function of estrogens was revived by the work of Villee et al. (1960) and Talalay and Williams-Ashmann (1960) on the stimulation of transhydrogenases by estradiol. Transhydrogenases transfer hydrogen between the $NAD^+/NADH$ and the $NADP^+/NADPH$ system. The idea was that estradiol is dehydrogenated with NAD^+ to yield estrone and NADH, and the estrone is then reduced again by NADPH to yield estradiol and $NADP^+$ (Fig. 2).

For a short time, there was some excitement about this hypothesis, especially in the USA, but this faded soon. Indeed, the transhydrogenation, even if it were true, would not explain any of the actions of the hormones. Moreover, Jensen and Jacobson (1962) could demonstrate that estradiol tritiated at C-17 does not lose tritium when administered at physiological doses to ovariectomized rats.

The concept of allosteric control of enzymes by allosteric effectors (Changeux 1961; Monod et al. 1963) was also applied to the mechanism of hormone action. Tomkins and Yielding (1961) discovered that glutamate dehydrogenase undergoes reversible dissociation into subunits when treated with steroids; but this effect is unspecific.

Finally, I would like to mention the hypothesis by Carl Martius (1955) presented here in Mosbach in 1954 that thyroxine may exert its action through uncoupling of oxidative phosphorylation. Though thyroxine is not a steroid hormone, such an effect could also be visualized for some steroid hormones which are known to increase the respiration of tissues in organ culture.

However, none of these hypotheses have been substantiated.

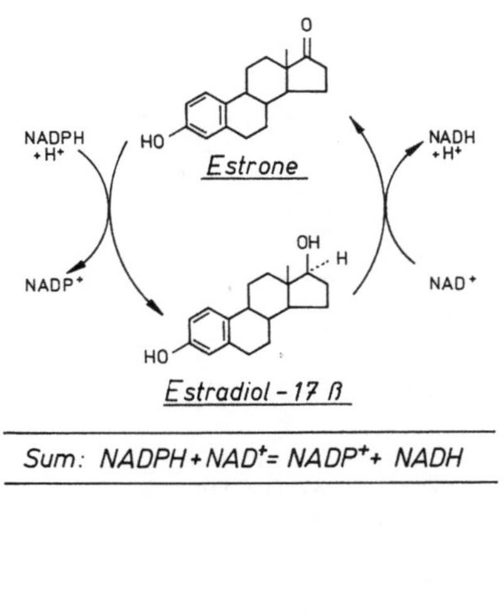

Fig. 2. Estrone/estradiol as "catalysts" in the transhydrogenation reaction

3 The Site of Action of Steroid Hormones

According to the concepts quoted above, the site of action of steroid hormones should be the cytosol or – alternatively – the mitochondrial matrix, where most of the metabolic reactions within the cell are located. Only in the hypothesis of uncoupling of oxidative phosphorylation, was the mitochondrial membrane involved. Some people looked at the cell membrane as a site of action and, indeed, around 1960 it was found that adrenaline and many peptide hormones interact with receptors on the cell membrane. But that is not my topic.

Today, we know that the main site of action of steroid hormones is the cell nucleus, more specifically the chromatin. This was not postulated on theoretical grounds, but it was revealed by an experiment: the induction of puffs on giant chromosomes of insect larvae by the steroid hormone ecdysone.

Ecdysone is the moulting and metamorphosis hormone of insects, which was isolated by Butenandt and Karlson (1954). Its structure is given in Fig. 3, together with the structure of 20-hydroxyecdysone; the latter is presumably the substance that is active in the target tissue.

The idea to use ecdysone to induce puffs in the salivary gland chromosomes of the larvae of the midge *Chironomos* arose through a conversation of Ulrich Clever and myself in Tübingen in the summer of 1959. Clever was interested in cultivating insect tissue and wanted to know whether ecdysone would be helpful as an additive in the culture medium. During this conversation, he told me about his observation of changes in the puffing pattern shortly before pupation. I immediately realized that this might be due to the endogenous production of ecdysteroids in the transition from the larva to prepupa and pupa and suggested a collaboration. This started in October, and within a couple of weeks we had positive results.

It should be stressed that puff induction by ecdysone is very sensitive, i.e., it can be induced by very low doses of ecdysone (0.1 pg mg^{-1} larvae). It is also very rapid; the

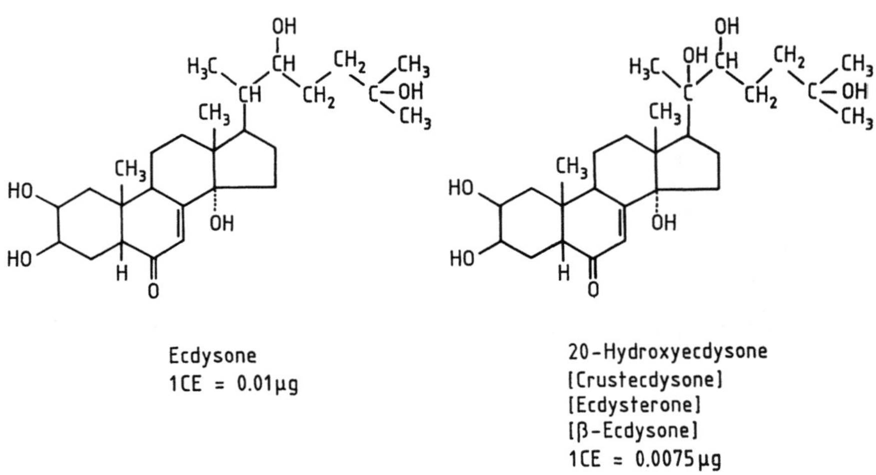

Fig. 3. Structural formulae of ecdysone and 20-hydroxy-ecdysone

effect can be seen within 30–60 min after injection; and it is highly specific: out of the about 10 000 visible chromosomal bands, only 2 react with puffing within the first 60 min. These three findings indicate that it is very probable that the activation of genes is the very first action of the ecdysteroid, not a later effect as has been postulated by Kroeger (1963).

Clever and I immediately realized that this observation was the clue to the mechanism of action of ecdysone, and in our first publication (Clever and Karlson 1960) we wrote:

"Deshalb erscheint die Hypothese gerechtfertigt, daß die primäre Wirkung des Hormons in der Beeinflussung der Aktivität eines oder mehrerer spezifischer Genloci besteht."[1]

Puffs are sites of gene activity (Pelling 1964). For the biologist, the idea that hormones are the tools to control gene activity is exciting, especially for morphogenetic hormones like ecdysteroids. In developmental biology, the idea was prevailing that development requires an ordered sequence of gene activities. If such hormones like the metamorphosis hormone could activate genes, it was understandable that a number of morphological and other changes occur at the time of pupation. The hormone was so to speak a "timing device" to control coordinate gene activity.

4 Biochemical Implications of Gene Activation

Around 1960, enough was known about molecular biology to draw a scheme (Fig. 4) of the mechanism of action of ecdysone which I published in early 1961 (Karlson 1961). It was postulated that "activation" of a gene, i.e., puff induction, leads to the synthesis of RNA which is then transferred to the cytosol and translated into protein, possibly an enzyme. We already had a good candidate at that time, namely, the enzyme converting protyrosinase into active tyrosinase. The latter part of the story was later replaced by the induction of a dopa decarboxylase which could indeed be demonstrated (Karlson and Sekeris 1962).

This first publication in a medical journal in the German language did not gain much publicity. However, a later review in *Perspectives in Biology and Medicine* (Karlson 1963) had some impact, partly because I could include work from other laboratories which corroborated the idea that steroid hormones act through induction of RNA and protein biosynthesis; it explained the induction of several enzymes by corticosteroids.

5 The Role of Steroid Receptors

A full understanding of the biochemical action of steroid hormones became only possible after the discovery of steroid receptors by Jensen and Jacobson (1962). These are intracellular proteins that bind the steroid hormone, and it is the receptor-hormone

[1] Italics in the original.

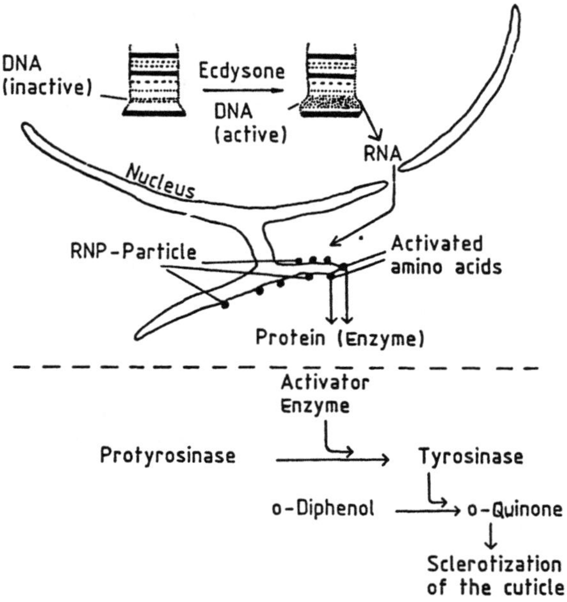

Fig. 4. Biochemical consequences of gene activation by ecdysone in insects. As a result of gene activation, RNA is produced and binds to ribosomes at the endoplasmic reticulum. There it is translated into protein, in this case an enzyme, which was believed, at that time, to be the activator enzyme of the tyrosinase. As a result, sclerotization of the larval cuticle occurs. (Karlson 1961)

complex that regulates transcription. We will hear more about these mechanisms in the course of this colloquium. At present, it is still controversial whether the receptor is present in the nuclear compartment or in the cytosol, but there is no doubt that the complex with the hormone acts at the nucleus. This is outlined in an updated scheme of the mechanism of action (Fig. 5).

The receptor is a protein which is present only in those tissues that react to the hormone, i.e., the target tissues. This explains why the action of hormone is limited to some tissues and is not a general effect in all cells.

6 On the Evolution of Steroid Hormones

In the last two decades, it became clear that the scheme developed for the action of ecdysone holds for all steroid hormones: The steroid hormone-receptor complex initiates transcription at certain genes, resulting in the production of pre-mRNA and eventually a protein. This is often an enzyme, but not necessarily: Some other proteins, e.g., ovalbumin, avidin, and uteroglobin, are induced by steroid hormones. Since most other hormones have a different mechanism of action, one may well ask why the mechanism of all steroid hormones is alike. My answer is that this is due to evolution (Karlson 1983, 1985).

This is not the place to repeat all that said already in my Butenandt Lecture 1983 cited above, but I would like to outline some essential points. The basic idea is that the hormonal systems using steroids as messengers evolved from an *intracellular* mechanism of the control of the activity of certain genes through *external* signals.

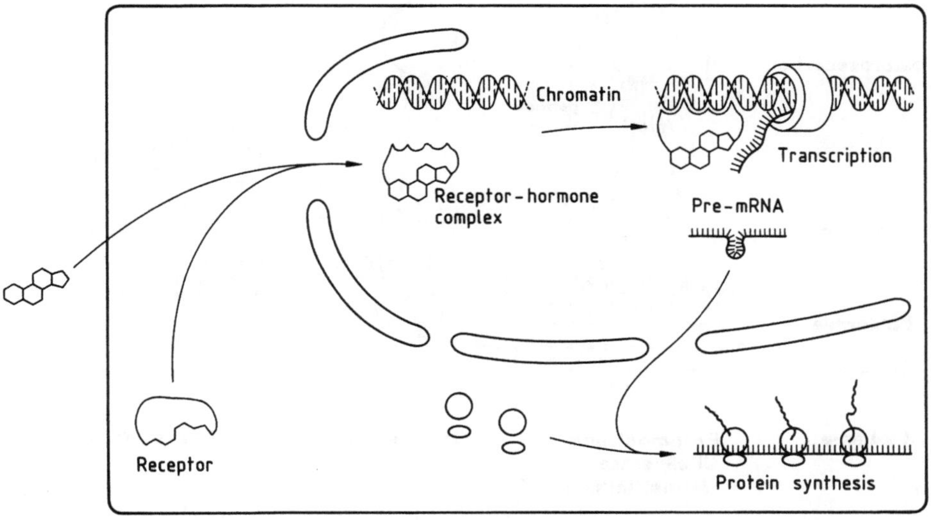

Fig. 5. Mechanism of action of steroid hormones. The steroid enters the cell, combines with the receptor protein, and the complex is transferred to the nucleus. There the receptor-hormone complex binds to a certain DNA sequence and initiates transcription. The pre-mRNA is processed and exported into the cytosol, where translation (protein synthesis) occurs. (Karlson 1988)

Ecdysteroids are important developmental hormones not only in insects but also in other phyla of invertebrates, in the Chelicerates and the Crustaceans. It is very interesting that in a very primitive arthropod, i.e., *Limulus*, the ecdysteroids are produced in a tissue which is also the target tissue. The external stimulus seems to be a neurohormone (Fig. 6a). In the Macrura, certain parts of the epidermis evolved into a tissue which is, in part, an "endocrine tissue" (Fig. 6b) and finally, in the Brachyura, an endocrine organ (the Y-organ) has developed from part of the epithelium of the epidermis (Fig. 6c).

This morphological evolution suggests that the functional evolution of this hormonal system was as follows: At first, the external signal (a neurohormone, presumably a peptide) acted at the level of the cell membrane and somehow activated a steroid hydroxylase that hydroxylated the cholesterol in the membrane (either in the inner layer of the cell membrane, or some membrane in the endoplasmic reticulum). The polyhydroxylated cholesterol leaves the membrane since it is hydrophilic; it is transferred to the nucleus and binds there to a protein which is the evolutionary precursor of the steroid hormone receptor. This complex then activates transcription of certain genes. In the next stage, the cells of the epidermis produce hydroxylated cholesterol not only as internal messenger but also as external messenger for neighboring cells in the epidermis. And finally (Fig. 6c) the tissue fully developed into a hormonal gland (the Y-organ) that produces polyhydroxylated cholesterol derivatives, i.e., ecdysteroids (in the case of Brachyura 20-hydroxy-ecdysone).

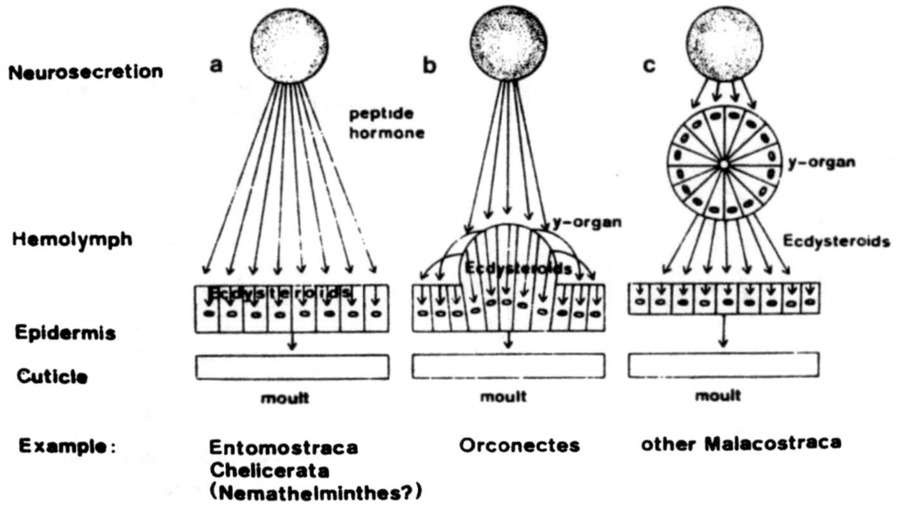

Fig. 6. Evolution of an ecdysteroid-producing gland in the arthropods. *a* The situation in *Limulus,* where ecdysteroids serve as intracellular second messengers. *b* Part of the epidermis evolves into a tissue which uses ecdysteroids not only as intracellular messenger, but it also excretes some of the ecdysteroids. *c* Part of the epidermis has evolved into the Y-organ, which is now a true ecdysteroid-producing gland. (Karlson 1985)

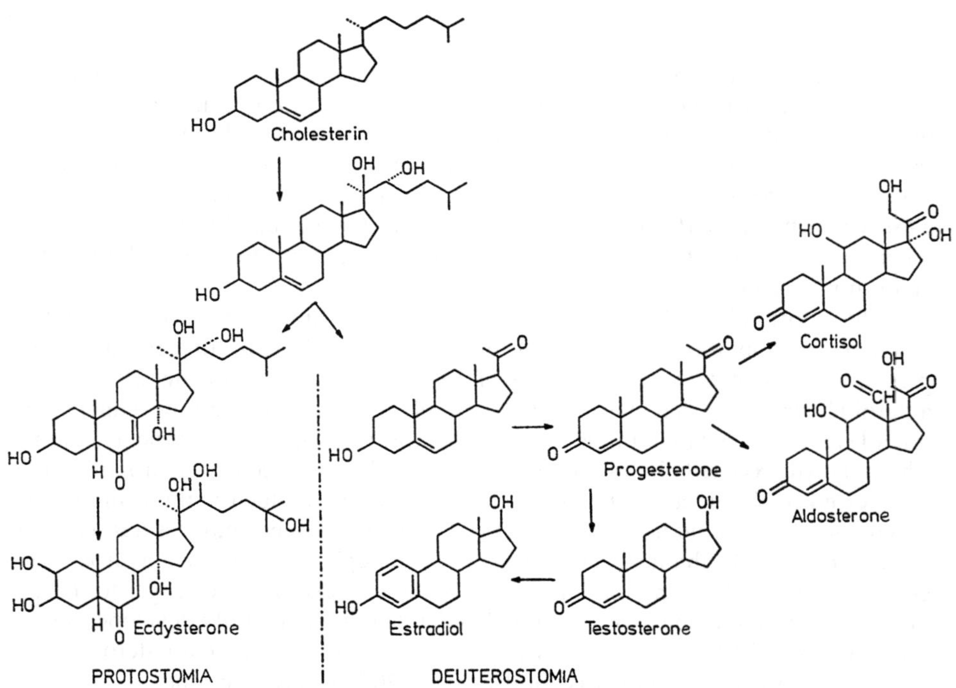

Fig. 7. Main lines of steroid hormone biosynthesis as evolutionary tree. (Karlson 1983)

Ecdysteroids seem to be the steroid hormones of all protostomia. In the deuterostomia, an important step in the evolution of steroid hormones has occurred: an enzyme appeared which cleaves the side chain so that cholesterol is converted into a pregnane derivative (pregnenolone). With the dehydrogenation of pregnenolone to progesterone, we have the first steroid hormone in the deuterostomia.

Indeed, one can read the transformation of cholesterol into the various steroid hormones as an evolutionary tree (Fig. 7). Progesterone, the most primitive steroid hormone of the deuterostomia, is converted first into adrenal steroids; further into testosterone; and finally into estradiol. With this scheme, we can understand why a hormone is simultaneously the precursor of the next hormone, and why the receptor proteins of steroid hormones are presumably homologous, since they have evolved from *one* primitive steroid binding protein that serves as receptor and control element in gene transcription.

This last section was not devoted to the history of our concepts of the mode of action of steroid hormones, but rather to recent developments that, I think, have helped to understand the hormonal systems using steroid molecules as signal substances.

References

Butenandt, A (1936) Ergebnisse und Probleme in der biochemischen Erforschung der Keimdrüsenhormone. Naturwissenschaften 24:529–536; 545–552

Butenandt, A & Karlson, P (1954) Über die Isolierung eines Metamorphose-Hormons der Insekten in kristallisierter Form. Z. Naturforsch 9b:389–391

Changeux, JP (1961) The feedback control mechanism of biosynthetic L-threonine deaminase by L-isoleucine. Cold Spring Harbor Symp Quant. Biol 26:313–318

Clever, U & Karlson, P (1960) Induktion von Puff-Veränderungen in den Speicheldrüsenchromosomen von Chironomus tentans durch Ecdyson. Exp Cell Res 20:623–626

Green, DE (1941) Enzymes and trace substances. Adv Enzymol. 1:177–198

Jensen, EV & Jacobson, HI (1962) Basic guides to the mechanism of estrogen action. Recent Progr Hormone Res. 18:387–414

Karlson, P (1961) Biochemische Wirkungsweise der Hormone. Dtsch Med Wochenschr 86:668–674

Karlson, P (1963) New concepts on the mode of action of hormones. Perspect Biol Med 6:203–214

Karlson, P (1983) Why are so many hormones steroids? (8. Adolf-Butenandt-Lecture). Hoppe-Seyler's Z. Physiol Chem 364:1067–1087

Karlson, P (1985) Warum sind so viele Hormone Steroide? (Gedenkvorlesung für Kurt Mothes). Nova Acta Leopoldina N.F. 57, 261:1–26

Karlson, P (1988) Kurzes Lehrbuch der Biochemie, 13th edn.

Karlson, P & Sekeris, CE (1962) Zum Tyrosinstoffwechsel der Insekten. IX. Kontrolle des Tyrosinstoffwechsels durch Ecdyson. Biochim Biophys Acta 63:489–495

Kroeger, H (1963) Chemical nature of the system controlling gene activation in insect cells. Nature (London) 200:1234–1235

Martius, C (1955) Die Wirkungsweise des Schilddrüsenhormons. In: 5. Kolloq. Ges. Physiol Chem., Mosbach. Springer, Berlin, pp 143–161

Monod, J, Changeux, JP & Jacob, F (1963) Allosteric proteins and cellular control systems. J Mol. Biol. 6:306–329

Pelling, C (1964) RNS-Synthese der Riesenchromosomen. Chromosoma (Berlin) 15:71–122

Talalay, P & Williams-Ashman, HP (1960) Participation of steroid hormones in the enzymatic transfer of hydrogen. Recent Progr Hormone Res. 16:1–47

Tomkins, GM & Yielding, KL (1961) Regulation of the enzymic activity of G glutamic dehydrogenase mediated by changes of its structure. Cold Spring Harbor Symp. Quant. Biol. 26:331–341

Villee, CA, Hagerman, DD & Joel, PB (1980) An enzymatic basic for the physiological function of estrogens. Recent Progr Hormone Res 16:49–77

Interaction of Steroid Hormone Receptors with DNA

M. Beato[1], U. Brüggemeier[1], G. Chalepakis[1], B. Gross[1], B. Piña[1], M. Schauer[1], E.P. Slater[1], and M. Truss[1]

1 Introduction

Modulation of gene expression frequently takes place at the level of transcription, and is mediated by the interaction of regulatory proteins with specific DNA sequences near the regulated promoters. Among the best-characterized examples of such regulatory proteins are the steroid hormone receptors. These proteins are members of a large family of nuclear proteins called the steroid/thyroid hormone receptor family or the nuclear receptor family. Most members of this gene family share a double specificity; on the one hand, they interact with small ligands, such as steroid hormones, thyroid hormones, vitamin D, and retinoic acid; on the other hand, they bind to regulatory elements in the DNA that mediate modulation of gene activity in *cis.* During the last decade the DNA complementary to the mRNAs encoding most of the members of the hormone receptor gene family have been cloned. A comparison of the primary structure of the various members of the family permits one to derive a general scheme for the modular architecture of these regulatory proteins (Fig. 1). The proteins are composed of three separate domains, one of which interacts with the hormone-ligand, the second binds to DNA, and the third domain modulates transcriptional activity. The most conserved region of the protein is located in the central domain, which consists of a short stretch of about 70 amino acids containing the DNA binding activity. This region of the protein exhibits an array of cysteine residues compatible with an organization of the domain into two so-called zinc fingers, in which two zinc ions are each tetrahedrally coordinated with four cysteines (for a review, see Evans 1988). The cysteine residues involved in binding to the zinc ions are probably those depicted in Fig. 1 for the glucocorticoid receptor, based on genetic analysis and on physical studies (Severne et al. 1988; Freedman et al. 1988). Mutational analysis along this region of the zinc fingers suggests that the amino terminal zinc finger is responsible for a specific DNA recognition, whereas the carboxy-terminal finger is only contributing to the total DNA affinity, but not to base recognition (Green et al. 1988; Evans 1988). In agreement with this interpretation, we found that the receptor covers four subsequent turns of the DNA double helix, but only the two central helical turns containing the conserved 15-mer DNA motif recognized by the receptor are contacted by the protein through the major groove of the double helix (Chalepakis et al. 1988). As the hormone receptors bind to the DNA in the form of a homodimer (Chalepakis et al. 1988b; Kumar and Chambon 1988; Fey et al. 1988), we believe that the two amino terminal zinc fingers

[1]Institut für Molekularbiologie und Tumorforschung, Emil-Mannkopff-Str. 2, D-3550 Marburg, Fed. Rep. of Germany

40. Colloquium Mosbach 1989
Molecular Mechanisms of Hormone Action

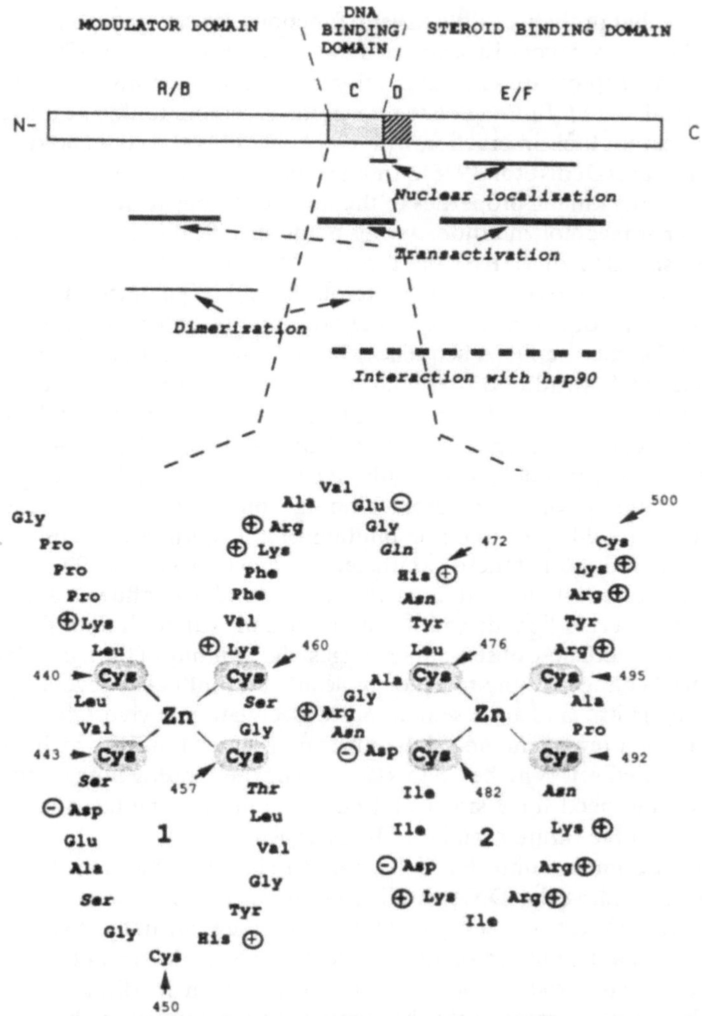

Fig. 1. Schematic representation of the linear structure of the nuclear receptors. The regions corresponding to functional activities are indicated. The lower part of the figure indicates the amino acid sequence of the DNA binding domain of the glucocorticoid receptor with the most probable coordination of the zinc ions

of the homodimer contact the DNA bases in the two halves of the conserved 15-mer, whereas the two carboxy-terminal zinc fingers interact with the backbone of the flanking nucleotides in the double helix. The exact identification of the amino acid residues involved in base recognition awaits the elucidation of the three-dimensional structure of the DNA-bound receptors.

Less well conserved among the different members of the nuclear receptor family is the carboxy-terminal half of the molecule that is known to interact with the small molecular weight ligands. This domain is encoded in many separate exons and has a complicated structure that probably fulfills several functions. For instance, binding of the hormone ligand seems to favor dissociation of the receptor from the hsp 90, a heat shock protein to which receptors are associated prior to hormone administration (Joab et al. 1988; Denis et al. 1988). Binding of the hormone receptors to this heat shock protein appears to prevent their interaction with DNA, possibly through masking of the DNA binding domain (Denis et al. 1988). Another consequence of hormone ligand binding is a change in the kinetic properties of the interaction between receptor and DNA. The steroid-free receptor in crude cytosol binds specifically to the hormone regulatory or responsive element (HRE), but the rates of association and dissociation are slow compared to those measured with the steroid-bound receptor (Schauer et al. 1989). This change in the kinetic behavior of the hormone-receptor complex could facilitate the search for specific DNA sequences in the context of a great excess of nonspecific genomic DNA. In addition, it seems that the dimerization of the receptor is a prerequisite for high affinity binding to the HRE and is also highly dependent on binding of the ligand (Kumar and Chambon 1988; Tsai et al. 1988). Finally, of course, transactivation of adjacent promoters is strictly dependent on ligand binding. Although mutant receptors lacking the steroid binding domain are able to activate adjacent genes, for the wild-type receptor binding of the hormone is absolutely required for gene activation. All of these parameters: interaction with hsp 90, kinetics of DNA-binding, dimerization, and transactivation are similarly influenced by the chemical nature of the steroid ligand; anti-hormones usually cannot induce the type of changes agonists initiate. These observations suggest that binding of the ligand leads to complex structural changes of the receptor molecule that influence several parameters, as measured in vitro, and are essential for transactivation in vivo. One should mention that there are a few members of the nuclear receptor family for which no ligand has been identified yet. Whether in this case the ligand binding domain fulfills its function without the need for a small molecular ligand, or whether additional ligands will be found in the future remains to be established.

The most variable domain within the nuclear receptor family is the amino terminal half of the protein preceding the DNA binding domain. In some members of the family, like the vitamin D receptor, this part of the molecule is virtually absent. This domain is also the most antigenic region of the receptor protein. When antibodies to the receptor are generated in rodents or in rabbits, most of them are directed against epitopes within the amino terminal variable region. Several functions have been mapped to this domain. In the glucocorticoid receptor, a transactivation domain has been located within the amino terminal region (Hollenberg and Evans 1988). In the case of the progesterone receptor, there is evidence suggesting that species of the receptor differing at the amino terminal end may specifically interact with different sets of genes (Tora et al. 1988). However, in many different assays, deletion of the amino terminal region does not completely inactivate the hormone receptor, and, therefore, it seems that this region functions as a modulator of the transactivational activity exerted by the remaining regions of the protein.

2 Results and Discussion

2.1 *The DNA Regulatory Elements*

A comparison of the DNA sequences recognized by the glucocorticoid receptor in many different target genes yields a consensus sequence of 15 base pairs (Fig. 2). This consensus sequence exhibits an imperfect palindromic structure centered at position 8. The right half of the palindrome is a highly conserved hexanucleotide motif, whereas the left half is less strictly conserved in different genes. Usually, more than one copy of this 15-mer is found in the hormone regulatory elements of different genes, and the different copies within a single gene are not highly conserved. In some cases, as in the mouse mammary tumor virus (MMTV) promoter, only parts of the palindromic structure are present in some of the repeats. From methylation protection experiments it is clear that the guanine residues in the two halves of the palindrome are important for recognition by the hormone receptor. Methylation of the N-7 position of these guanines interferes with receptor binding (Scheidereit and Beato 1984). Because the two blocks of guanine residues that are contacted in the 15-mer are separated by one turn of the double helix, we postulated that a dimer of the receptor with a diad symmetry axis interacts with the conserved DNA element (Scheidereit and Beato 1984; Scheidereit et al. 1986). This idea has been supported by experimental data and is now generally accepted (Kumar and Chambon 1988; Tsai et al. 1988).

In addition to the contacted guanine residues, we know that there are intimate contacts between the hormone receptors and the thymine residues at position 10 and 12 in the upper strand, and at position 4 in the lower strand (Fig. 2). Weak contacts are also observed with the thymines at positions 15 in the upper strand and 6 in the lower strand. These contacts to the thymine residues can be detected by the binding interference observed after modification of the thymine bases with low concentrations of potassium permanganate, which oxidizes the C5-C6 double bond and opens the pyrimidine ring at this position (Howgate et al. 1968; Rubin and Schmid 1980; Truss and Chalepakis unpublished). Of particular interest is the contact to the T at position 12, as this position differs between the HREs responding to the two subfamilies of nuclear receptors (Beato 1989). Whereas the HREs for glucocorticoids, progesterone, androgens, and mineralocorticoids exhibit a T at this position, those for estrogens, retinoic acid, and thyroid hormones contain a conserved A. It is conceivable that the different position of the 5-methyl group in the major groove serves to distinguish these two subtypes of HREs.

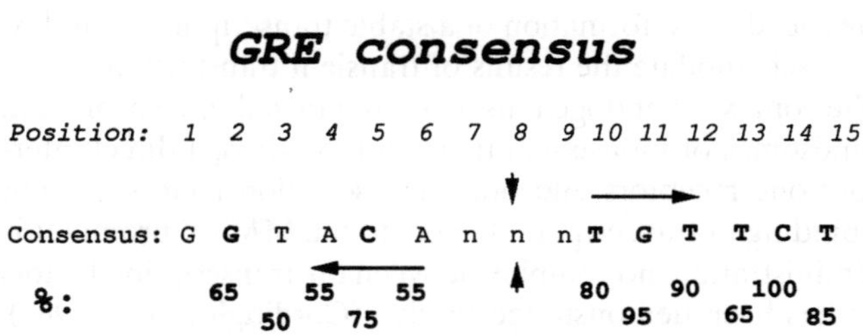

Fig. 2. Sequence of the conserved 15-mer recognized by the glucocorticoid receptor. Only the sense strand is shown. The *numbers* below the sequence indicate the percentage of binding sites (n = 20) in which this particular nucleotide is found

Although only 15 base pairs are conserved in the different receptor binding sites, it is clear that the DNA sequences flanking the 15-mer are important for the DNA affinity of the receptor. In fact, oligonucleotides containing a 15-mer and only two or three flanking base pairs, exhibit a very low affinity for the hormone receptor in vitro. As additional base pairs are added at each end of the oligonucleotide, the affinity for the receptors increases and reaches an optimal value when ten base pairs are added to each side of the 15-mer (Chalepakis unpublished). Since the sequences flanking the 15-mer are not conserved between different receptor binding sites, it seems that the flanking sequences contribute to receptor affinity by a nonspecific interaction. According to a recently proposed model, an interaction of the carboxy-terminal finger of the DNA binding domain – that does not contribute to specific base recognition – with the backbone of the double helix would account for the observed effect of the DNA flanking sequences (Chalepakis et al. 1988b).

When several copies of the 15-mer are present in a gene regulatory element, they usually act in a cooperative way, although there is no evidence for a strong cooperativity of receptor binding to linear DNA (Chalepakis et al. 1988a). It seems that after the individual proteins have bound to their respective sites, interactions between the receptor molecules and between the receptors and other transcription factors, are important for transactivation. This view is supported by the observed influence of DNA topology on both receptor binding and gene activation in transfection experiments. When the binding sites for the receptors are separated on the DNA, and the HREs are moved away from the promoter, transcriptional activation by hormones is highly dependent on the DNA topology of the transfected plasmids (B. Piña unpublished). While negatively supercoiled plasmids respond very efficiently to progesterone administration, linearized DNA molecules are much less sensitive to hormonal induction. In parallel with these findings we observed a preferential binding of the hormone receptors to negatively supercoiled forms of plasmids containing the HREs (M. Truss unpublished).

2.2 Mechanism of Transactivation

In transient gene transfer experiments it has been shown that the binding sites for the receptor can activate transcription when located in the proximity of binding sites for other transcription factors (Ankenbauer et al. 1988; Schüle et al. 1988). According to a widespread model, this finding is interpreted as indicative of a direct interaction between the receptor bound to the HRE and transcription factors bound to the corresponding adjacent sites, leading to formation of a stable transcription complex. Though this model seems to accomodate the results of transient transfection experiments, it is less useful in the context of endogenous genes organized into chromatin. In particular, in minichromosomes of mouse mammary tumor virus, a direct interaction between steroid hormone receptors and other transcription factors has not been observed. Although binding of transcription factors to the MMTV promoter is dependent on hormone administration, no complex between the transcription factors and the hormone receptor has been demonstrated in vivo (Cordingley et al. 1987). There are possible explanations for this finding based on differential affinity of steroid hormone receptors and nuclear factor I for the respective sites, but another possible interpretation is that a complex including hormone receptors and nuclear factor I on

DNA is not formed in vivo. In vitro, binding of the steroid hormone receptors to the MMTV-HRE interferes with binding of nuclear factor I (NFI) to its cognate sequences on naked DNA, and vice versa (U. Brüggemeier unpublished). This competition between receptors and NFI is observed in vitro independently of DNA topology and no indication for a cooperativity is found. Thus, alternative mechanisms should be considered for explaining the hormonal regulation of the MMTV promoter.

2.3 Role of Chromatin Structure

It has been shown that the MMTV promoter region is precisely organized into nucleosomes in vivo, with a nucleosome-like structure covering the binding sites for the hormone receptors and for NFI (Richard-Foy and Hager 1987). After hormone administration the chromatin structure in this area of the promoter changes and the whole region becomes more accessible to DNase I (Zaret and Yamamoto 1984; Cordingley et al. 1987). Since hormone administration also leads to binding of NFI and a TATA-box binding factor, one could extrapolate that the function of the hormone-receptor complex is to remove a nucleosome-like structure in order to make the region accessible to transcription factors. According to this model, the promoter will be repressed normally by a precisely positioned nucleosome that masks the binding site for the transcription factors in the promoter. The function of the regulatory proteins such as the steroid hormone receptors will then be to change the structure of the chromatin in such a way that these sites become exposed for transcription factor binding. This attractive model is amenable to experimental testing, which essentially will consist in demonstrating that binding of the receptor really changes the structure of the nucleosome and makes it accessible to NFI. These experiments can now be performed in vitro since the MMTV promoter, when recombined with histone octamers in vitro, has been shown to form a precise nucleosome with a very similar positioning as described in vivo (Perlman and Wrange 1988). We have been able to reproduce these results with slightly different DNA fragments, and have shown that the receptor binds to the nucleosomally organized MMTV promoter with an affinity only four–five fold lower than in naked DNA (B. Piña and U. Brüggemeier unpublished). On the other hand, NFI binds to the nucleosomally organized MMTV regulatory region with at least 50-fold lower affinity than to naked DNA. These findings are compatible with the idea that the nucleosome located over the HRE is a repressor of transcription from the MMTV promoter because it prevents binding of transcription factors, and that the receptors are still able to bind to the nucleosomally organized promoter. Ultimately, it will be necessary to demonstrate that the in vitro reconstituted nucleosomes are able to reproduce the behavior of the in vivo chromatin in a cell-free transcription assay.

3 Summary and Outlook

The binding of hormone receptors to the regulatory sequences called hormone responsive elements (HRE) has been clearly demonstrated to be a crucial step in hormonal induction of transcription. Knowledge of the details of this interaction, including the identification of regions within the receptor protein involved in base recognition and DNA binding, is progressing rapidly. However, the exact mechanism of transcriptional control is still not understood. We will need additional experimental assays to investigate the steps following binding of the receptor to the HREs. In addition to the development of cell-free transcription systems that would allow the dissection of the individual steps involved in gene regulation (Corthesy et al. 1988), many important questions still remain open and will have to be addressed in the near future.

Very little is known about the biosynthesis and the metabolic changes that modulate the activity of the hormone receptors. There are many indications for the existance of inactive forms of the receptors that are either unable to bind the hormone or to bind to DNA. How these activities are modulated is practically unknown. Phosphorylation could play a role (Auricchio 1989), but other covalent modifications of the hormone receptors have not been explored. Particularly interesting is the participation of other macromolecules in modulating the activity of the hormone receptors. In addition to the heat shock protein, hsp90, mentioned above, other protein components and even RNA have been detected in the heteromeric complexes containing unoccupied hormone receptors. In fact, even a low molecular weight modulator that fulfills some of the functions accomplished by exogenous molybdate, has been postulated (Bodine and Litwack 1988). The exact function of these factors and how they interact with the receptors is not yet known.

In addition to their known function as inducers of specific gene expression, steroid hormones are known to repress the transcription of a variety of genes in different cells (Beato 1989). The exact mechanism of transcriptional repression by steroid receptors is not known, though in some cases they seem to act by competing with other positive regulatory proteins or transcription factors (Akerbloom et al. 1988). Alternatively, other functional domains of the receptor protein distinct from those involved in gene activation could be responsible for repression of transcription. Future experiments will have to clarify the general validity of this mechanism.

Perhaps, the most relevant question to be answered in the future concerns the mechanism responsible for differential gene activation. Formulated in a simple way this question is as follows: Why do the same hormones acting through the same receptor proteins activate (or repress) different genes in different cells? In some cases, trivial explanations, such as the equipment of different cells with the different sets of transcription factors and regulatory proteins, will account for the specificity. In other cases, however, the history of the particular cell, its differentiation pathway, seems to set the stage for regulation of a restricted fraction of the cellular genome. How this differentiation history of each cell is fixed structurally and transmitted to the daughter cells is unknown, but many observations point to a participation of chromatin structure in this process. Unraveling the molecular basis of differential chromatin organization is one of the major challenges facing research on the regulation of gene activity in animal cells.

Acknowledgments. The experimental work reported in this paper has been supported by grants from the Deutsche Forschungsgemeinschaft and the Fonds der Chemischen Industrie.

References

Akerblom, I.E., Slater, E.P., Beato, M., Baxter, J.D. & Mellon, P. (1988) Negative regulation by glucocorticoids through interference with a cAMP-responsive enhancer. Science 241:350–353

Ankenbauer, W., Strähle, U. & Schütz, G. (1988) Synergistic action of glucocorticoid and estradiol responsive elements. Proc. Natl. Acad. Sci. USA 85:7526–7530

Auricchio, F. (1989) Phosphorylation of steroid receptors. J. Steroid. Biochem 32:613–622

Beato, M. (1989) Gene regulation by steroid hormones. Cell 56:335–344

Bodine, P.V. & Litwack, G. (1988) Purification and structural analysis of the modulator of the glucocorticoid-receptor complex. Evidence that the modulator is a novel phosphoglyceride. J. Biol. Chem. 263:3501–3505

Chalepakis, G., Arnemann, J., Slater, E.P., Brüller, H.-J., Gross, B. & Beato, M. (1988a) Differential gene activation by glucocorticoids and progestins through the hormone regulatory element of mouse mammary tumor virus. Cell 53:371–382

Chalepakis, G., Postma, J.P.M. & Beato, M. (1988b) A model for hormone receptor binding to the mouse mammary tumour virus regulatory element based on hydroxyl radical footprinting. Nucl. Acid Res. 16:10237–10247

Cordingley, M.G., Riegel, A.T. & Hager, G.L. (1987) Steroid-dependent interaction of transcription factors with the inducible promoter of mouse mammary tumour virus in vivo. Cell 48:261–270

Corthesy, B., Hipskind, R., Theulaz, I. & Whali, W. (1988) Estrogen-dependent in vitro transcription from the vitellogenin promoter in liver nuclear extracts. Science 239:1137–1139

Denis, M., Poellinger, L., Wikström, A.C. & Gustafsson, J.A. (1988) Requirement of hormone for thermal activation of the glucocorticoid receptor to a DNA-binding state. Nature (London) 333:686–688

Evans, R.M. (1988) The steroid and thyroid hormone receptor superfamily. Science 240:889–895

Freedman, L.P., Luisi, B.F., Korszun, Z.R., Basavappa, R., Sigler, P.B. & Yamamoto, K.R. (1988) The function and structure of the metal coordination sites within the glucocorticoid receptor DNA binding domain. Nature (London) 334:543–546

Green, S., Kumar, V., Theulaz, I., Wahli, W. & Chambon, P. (1988) The N-terminal DNA-binding "zinc-finger" of the oestrogen and glucocorticoid receptors determines target gene specificity. EMBO J. 7:3037–3044

Hollenberg, S.M. & Evans, R.M. (1988) Multiple and cooperative transactivation domains of the human glucocorticoid receptor. Cell 55:899–906

Howgate, P., Jones, A.S. & Tittensor, J.R. (1968) The permangatate oxidation of thymidine. J. Chem. Soc (C) 275–279

Joab, I., Radanyi, C., Renoir, M., Buchou, T., Catelli, M.G., Binart, N., Mester, J. & Baulieu, E.E. (1988) Common non-hormone binding component in the non-transformed chick oviduct receptors of four steroid hormones. Nature (London) 308:850–853

Kumar, V. & Chambon, P. (1988) The estrogen receptor binds tightly to its responsive element as a ligand-induced homodimer. Cell 55:145–156

Perlman, T. & Wrange, Ö. (1988) Specific glucocorticoid receptor binding to DNA reconstituted in a nucleosome. EMBO J. 7:3073–3079

Richard-Foy, H. & Hager, G.L. (1987) Sequence specific positioning of nucleosomes over the steroid-inducible MMTV promoter. EMBO J. 6:3221–2328

Rubin, C.M. & Schmid, C.W. (1980) Pyrimidine-specific chemical reactions useful for DNA-sequencing. Nucl. Acid Res. 8:4613–4619

Schauer, M., Chalepakis, G., Willmann, T. & Beato, M. (1989) Binding of hormone accelerates the kinetics of glucocorticoid and progesterone receptor binding to DNA. Proc. Natl. Acad. Sci. USA 86:1123–1127

Scheidereit, C. & Beato, M. (1984) Contacts between receptor and DNA double helix within a glucocorticoid regulatory element of mouse mammary tumor. Proc. Natl. Acad. Sci. USA 81:3029–3033

Scheidereit, C., Westphal, H.M., Carlson, C., Bosshard, H. & Beato, M. (1986) Molecular model of the interaction between the glucocorticoid receptor and the regulatory element of inducible genes. DNA 5:383–391

Schüle, R., Müller, M., Otsuka-Murakami, H. & Renkawitz, R. (1988) Many transcription factors interact synergistically with steroid receptors. Science 242:1418–1420

Severne, Y., Wieland, S., Schaffner, W. & Rusconi, S. (1988) Metal binding "finger" structures in the glucocorticoid receptor defined by site-directed mutagenesis. EMBO J. 7:2503–2508

Tora, L., Gronemeyer, H., Turcotte, B., Gaub, M.P. & Chambon, P. (1988) The N-terminal region of the chicken progesterone receptor specific target gene activation. Nature (London) 333:185–188

Tsai, S.Y., Carlstedt-Duke, J., Weigel, N.L., Dahlman, K., Gustafsson, J.A., Tsai, M.J. & O'Malley, B.W. (1988) Molecular interactions of steroid hormone receptor with its enhancer element: evidence for receptor dimer formation. Cell 55:361–369

Zaret, K.S. & Yamamoto, K.R. (1984) Reversible and persistent changes in chromatin structure accompanying activation of a glucocorticoid-dependent enhancer element. Cell 38:29–38

Clustered Arrangement and Interaction of Steroid Hormone Receptors with Other Transcription Factors

R. RENKAWITZ[1], R. SCHÜLE[1,2], C. KALTSCHMIDT[1], C. BANIAHMAD[1], A. BANIAHMAD[1], J. ALTSCHMIED[1,2], CH. STEINER[1], and M. MULLER[1]

1 Introduction

One of the main contemporary challenges in molecular genetics is the elucidation of the mechanisms underlying gene regulation. Several tissue-specific promoter or enhancer elements have been identified. Many of the model systems analyzed are limited in that only mature cells or tissues are available for testing, whereas gene regulation may occur in a multistep process during differentiation from precursor to mature cells and may be modulated at certain levels of differentiation by different hormones. The hematopoetic cell population is generated by one of the few differentiation pathways which can be studied step by step in vitro as well as in vivo. Expression of the lysozyme gene is a specific differentiation marker, being gradually turned on during maturation of macrophages. In addition to this constitutive macrophage expression, the chicken lysozyme gene is inducible by steroid hormones in the tubular gland cells of the oviduct.

2 Results and Discussion

2.1 Additive Activity of Several Chicken Lysozyme Regulatory Elements

Functional analysis of the chicken lysozyme gene has revealed a complex pattern of different regulatory sequences upstream of the transcriptional start site. Two hormone-responsive elements have been identified, one at -200 bp upstream of the start site, mediating progesterone as well as glucocorticoid induction in the oviduct (Renkawitz et al. 1984) and which can be bound by both steroid receptors in vitro (Renkawitz et al. 1984; von der Ahe et al. 1985). The other steroid-responsive element is inducible by progesterone and has been analyzed in human mammary carcinoma cells (Hecht et al. 1988). In addition to these hormone-responsive elements, there are several so-called silencer and enhancer elements (Fig. 1). Within 3000 bp upstream of the transcriptional start site are two enhancer elements (E-0.2 and E-2.7) and three silencer elements (S-2.4, S-1.0, and S-0.25) (Steiner et al. 1987; Baniahmad et al. 1987). We have tested these silencer and enhancer elements separately by combining individual sequences with a promoter/reporter gene fusion and transfecting these constructs into different recipient cell types to test their effect on the transcription of the

[1]Max-Planck-Institut für Biochemie, Genzentrum, D-8033 Martinsried, Fed. Rep. of Germany
[2]Present address: Salk Institute for Biological Studies, La Jolla, CA 92037, USA

40. Colloquium Mosbach 1989
Molecular Mechanisms of Hormone Action

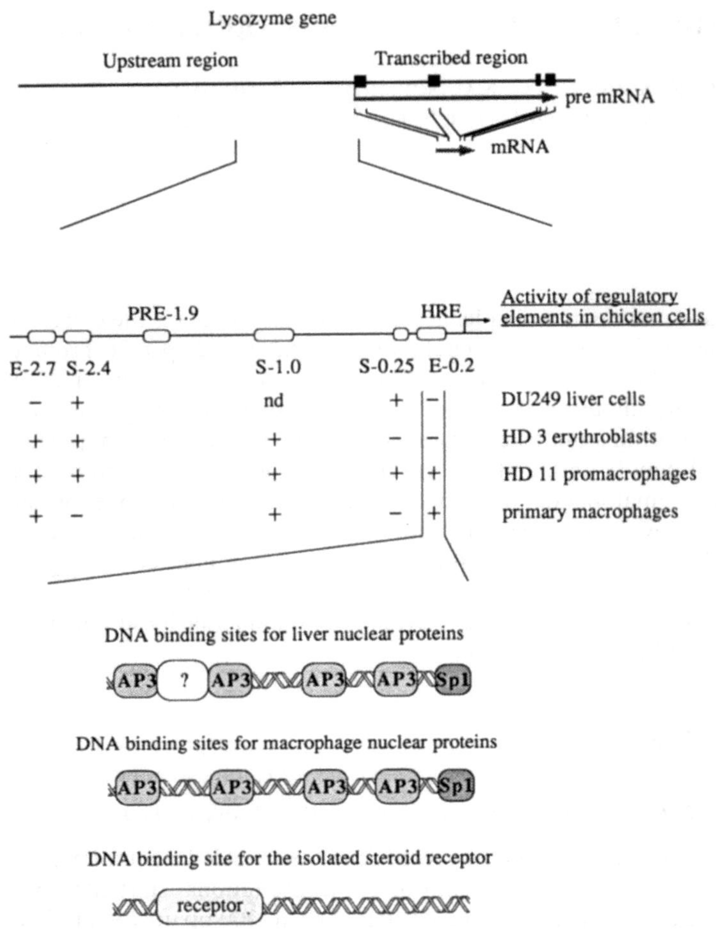

E-2.7	S-2.4	S-1.0	S-0.25	E-0.2	
−	+	nd	+	−	DU249 liver cells
+	+	+	−	−	HD 3 erythroblasts
+	+	+	+	+	HD 11 promacrophages
+	−	+	−	+	primary macrophages

Fig. 1. Regulatory activities of the chicken lysozyme gene. Gene structure and function are demonstrated on three levels. The upper level (*top* part of figure) depicts lysozyme exon sequences (*black boxes*) with intron and upstream sequences (*thin line*). The splicing event from pre-mRNA to mature mRNA is indicated. The second level (*central* part of figure) shows positions and activities of enhancer and silencer elements. (+) indicates changed activity of the tkCAT reporter gene by either enhancer (increased activity) or silencer (decreased activity) elements; (—) indicates unchanged activity of tkCAT. The fine structure level (*bottom* part of figure) illustrates for the E-0.2 kb enhancer the potential binding of different nuclear factors (see text) to enhancer modules

reporter gene. We chose four types of chicken cells: (1) a liver cell line, which does not express the endogenous lysozyme gene and does not belong to the hematopoetic pathway; (2) an erythroblast line, which does belong to the hematopoetic pathway, and which does not express the endogenous lysozyme gene; (3) promacrophages which express the endogenous lysozyme gene at an intermediary level; and (4) mature primary macrophages which show high expression of the endogenous lysozyme gene. Except for the silencer sequence at –1.0 kb all other elements show a tissue-specific activity (Fig. 1). Liver cells do not activate either enhancer element, but show repressing activity of both of the flanking silencer elements (S-0.25 and S-2.4). In erythroblasts the distal pair of the enhancer/silencer is active, whereas the proximal pair is inactive. Promacrophages show activity for all of the elements analyzed and mature primary macrophages show full activity of both enhancer elements with the adjacent silencer elements being inactive (Fig. 1).

This pattern of activity of different regulatory elements suggests that the additive effect of all elements controls lysozyme gene activity. Like in promacrophages with all elements being active only an intermediary gene activity can be achieved, whereas in erythroblasts silencer activity overrides the activity of the E-2.7 kb enhancer, which shows anyhow only a weak activity as compared to the positive cell types. To test this possible additive effect we used the lysozyme promoter which had been fused to the bacterial gene coding for chloramphenicol acetyl transferase (CAT) in order to allow efficient analysis of activity of the transfected gene. To this construct we added one or the other regulatory element (Fig. 2). Activity achieved in promacrophage cells with the E-0.2 construct was set to 100%. Deletion of 10 bp destroys the enhancer and leads to only a residual activity of 3%, whereas the addition of the second enhancer leads to an activity of 700%. The addition of one of the silencer elements either in its natural distance and configuration or separated by a 27 bp linker sequence from the enhancer leads in either orientation to about 20% CAT activity. Dimerization of the S-0.25 silencer leads to pronounced reduction in CAT activity. These data suggest that the additive effect of several regulatory elements which show different specificities together controls gene activity.

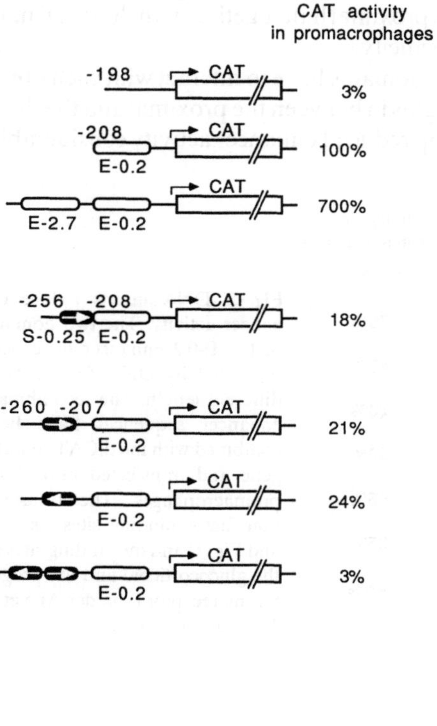

Fig. 2. Additive effect of silencer and enhancer elements. Several DNA recombinants were generated based on the lysozyme promoter fused to the CAT gene. Deletion mutant –208 contains the functional enhancer E-0.2 and S-0.25 is the silencer element upstream of E-0.2. Deletion mutant –256 contains the silencer element in its original position, whereas the other silencer containing constructs have been generated by insertion of a synthetic silencer sequence upstream of the E-0.2 enhancer. CAT activities have been measured after transfer of the respective DNA constructs into HD11 promacrophage cells

2.2 *Synergistic Activity of Enhancer Modules*

In order to identify transcription factors binding these regulatory elements we analyzed the nuclear proteins from different chicken cell types for their binding to specific sequences within silencer or enhancer elements. Figure 1 shows the results of such experiments with the E-0.2 enhancer. We found multiple binding sites within a DNA fragment of 150 bp. The binding specificities at four of these binding sites are similar to the known AP3 transcription factor. In addition to these binding sites, there is one site which is recognized by the Sp1 transcription factor. Basically identical binding patterns were achieved with nuclear proteins isolated from liver cells as well as from macrophages. With one exception there is an additional binding site for a liver protein in between the distal two AP3 binding sites. Whether the presence of this protein is connected with enhancer inactivity remains to be shown. Having found multiple binding sites for similar nuclear factors we wanted to know whether or not all of these sites are required for function and whether interaction between sites can be detected. We cloned several enhancer subfragments in various orientations and combinations upstream of the thymidine kinase (tk) promoter, which in turn was fused to the CAT gene (data not shown). It turned out that the whole length of the enhancer fragment with all of its protein binding sites was required for optimal enhancer function. Nevertheless, the enhancer could be rearranged such that the enhancer proximal and distal halves were joined together in a head-to-head or tail-to-tail orientation as compared to the wild-type arrangement. Even these combinations showed almost wild-type enhancer activity. Duplication of the distal two AP3 sites led to wild-type activity within promacrophage cells but lost the specificity by leading to some enhancer activity in liver cells. Duplication of the proximal two AP3 sites with the adjacent Sp1 site did not show any activity. These results (data not shown) suggest that only the combination of both enhancer domains (the proximal half together with the distal half) lead to enhancer activity and enhancer specificity.

To analyze whether these two enhancer domains have to interact with each other, we inserted DNA fragments of different lengths in between the proximal and the distal half of the enhancer (Fig. 3). Insertion of 5 bp reduced enhancer activity considerably,

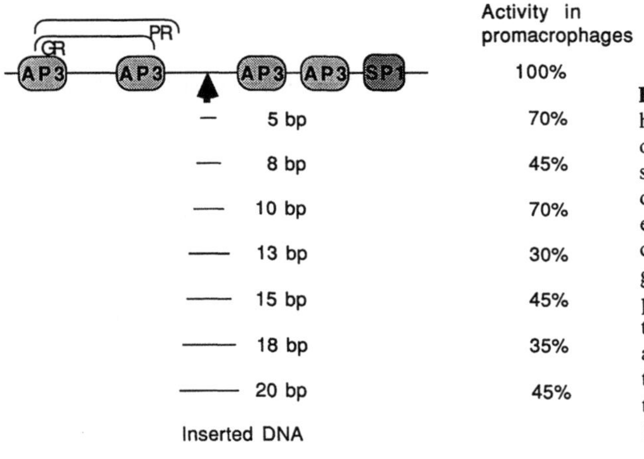

Fig. 3. DNA inserts reduce enhancer activity. The two domains of the E-0.2 enhancer have been separated by DNA fragments of different length. The recombinant enhancer sequences have been combined with the tkCAT reporter gene and transfected into HD11 promacrophages. The transcription factor binding sites (see text and Fig. 1) and the binding sites for the glucocorticoid and the progesterone receptor (von der Ahe et al. 1985) are indicated

with a further reduction caused by a 8 bp spacer. However, a spacer of 10 bp led to an increase in enhancer activity, presumably allowing the proteins bound to the two enhancer halves to be positioned in a three-dimensional arrangement similar to that in the wild-type enhancer. Further separation of the elements again resulted in reduced activities. This result indicates the requirement for a certain stereo-specific alignment of the enhancer domains in order to function optimally. Therefore, an interaction of the bound proteins on the two enhancer domains is suggested, which may be a direct interaction or may involve a "bridging protein" mediating the interaction of DNA-bound factors. A first example of an enhancer subdomain interaction was that of the SV40 enhancer (Takahashi et al. 1986), although direct biochemical proof for such an interaction has still not been achieved.

2.3 Synergistic Activity of Steroid Receptor and Other Transcription Factors

In addition to the observed cell type specific enhancer activity, the E-0.2 enhancer is inducible in chicken oviduct cells by two steroids: glucocorticoids and progestins (Renkawitz et al. 1984). Footprinting experiments with both isolated receptors revealed a binding to the distal half of the enhancer (von der Ahe et al. 1985) as indicated in Fig. 1. The receptor binding site seems to be embedded in several binding sites for other transcription factors. Similar clustered arrangements of receptor and transcription factor binding sites have been seen in other systems, too (Speck and Baltimore 1987; Cato et al. 1988; Danesch et al. 1987; Lee et al. 1987; Miksicek et al. 1987). Therefore, we wanted to know whether transcription factors binding adjacent to receptor binding sites might influence the steroid response. We combined a well-characterized receptor binding site (MMTV sequence −186/−170, which has been shown to bind progesterone as well as glucocorticoid receptors; Payvar et al. 1983; Scheidereit et al. 1983; Strähle et al. 1987), with well-characterized binding sites for other transcription factors (Fig. 4). The factor binding sites were the functional consensus sequence of NF1 binding sites (Miksicek et al. 1987), the CAAT box [recognized by the CAAT-binding protein CBP (Chodosh et al. 1988)], the Sp1 binding site (Barrera-Saldana et al. 1985), the CACCC box as found in the tryptophan oxygenase gene (Danesch et al. 1987; Schüle et al. 1988a), the binding site for the octamer transcription factor (Rosales et al. 1987), and a second receptor binding site. We inserted these combinations upstream of a tkCAT reporter gene. Transfection of these recombinant plasmids into a progesterone receptor containing cell line led to a very high induction by progesterone (Schüle et al. 1988b) as compared to the inducibility seen with a control plasmid, which has only the receptor binding site (PRE) and a mutagenized CACCC sequence (CACCC*) unable to bind the CACCC-nuclear factor. Clearly the presence of a functional transcription factor binding site leads to a synergistic effect on progesterone induction (Fig. 4). Similar effects were achieved with the glucocorticoid receptor binding to the same MMTV receptor binding site (Schüle et al. 1988a).

To investigate the importance of the arrangement of the different elements, we constructed and tested recombinants which contain the CACCC box between GRE and tk promoter in either sense or antisense orientation, or which contain two CACCC sequences (Fig. 5). Glucocorticoid induction revealed that these plasmids are all equally well inducible. These results suggest that the CACCC box/GRE unit works

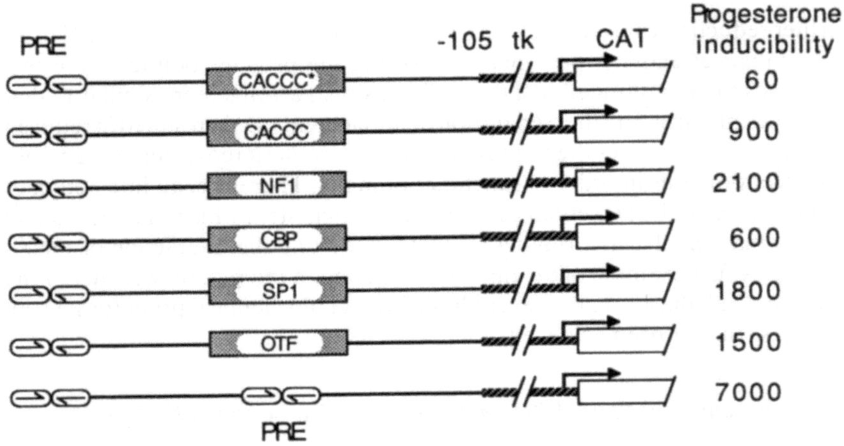

Fig. 4. Synergistic activity of the progesterone receptor with adjacent transcription factors. The MMTV progesterone responsive element (*PRE*) was combined with binding sites for another progesterone receptor, for *OTF, Sp1, CBP, NF1,* or the *CACCC* box (for details, see text). For reference, a mutagenized CACCC element (*CACCC**), which cannot bind a protein, was combined with the PRE. After fusion to the tkCAT reporter gene and transfection into T47D breast cancer cells the inducibility by progesterone was determined

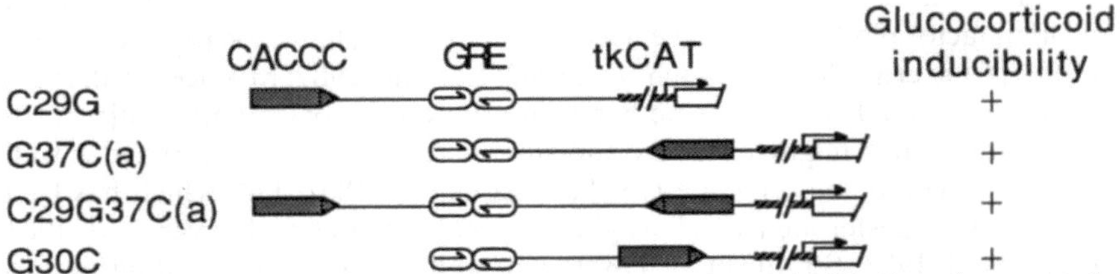

Fig. 5. Different arrangements of CACCC box and GRE are equally inducible. Several constructs fused to the tkCAT reporter gene are schematically drawn. The position of the CACCC element is indicated by the *shaded boxes* and its orientation is shown by the *arrowheads*. Glucocorticoid induction was measured in mouse L cell fibroblasts (for details, see Schüle et al. 1988a)

independently of its orientation relative to the promoter and that the presence of a second CACCC box does not further increase the inducibility. Whether these results suggest a direct interaction between CACCC box protein and the receptor and whether only one pair of factors can interact at a time have to be shown biochemically. That there is interaction between a transcription factor and a steroid receptor either directly or indirectly via a "bridging protein" is suggested by the following experiment. We inserted DNA of different lengths between the CACCC sequence and the GRE/PRE upstream of tkCAT. Inducibility by the glucocorticoid dexamethasone or by the progestin R5020 shows a cyclic pattern: Several peaks are apparent, corresponding to spacings of approximately 10 bp (Fig. 6). This result may best be explained by the requirement for a stereo-specific alignment of the transcription factors involved. This interpretation is substantiated by the fact that the two different steroid receptors with their different molecular weights and different binding geometries require different stereo-specific alignments as judged by the two different curves (Fig. 6) obtained with dexamethasone or R5020 induction.

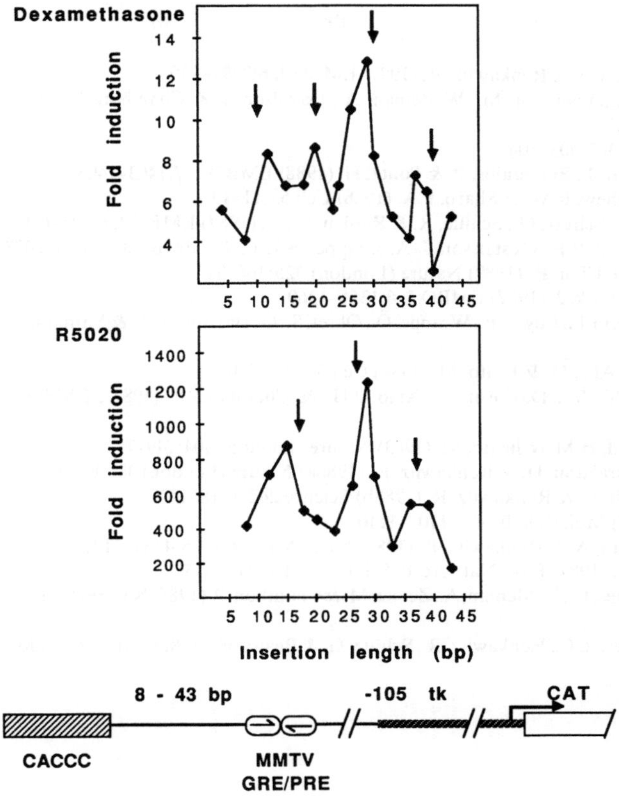

Fig. 6. Steroid-specific space requirement for the GRE/PRE and the CACCC element. DNA fragments of different lengths have been inserted between the receptor binding site (GRE/PRE) and the CACCC element. Inducibility of these constructs upstream of the tkCAT reporter gene by glucocorticoid dexamethasone was tested in mouse L cells, and inducibility by progestin R5020 was determined in T47D breast cancer cells. The *arrows* point to a 10 bp periodicity

3 Summary and Outlook

Analysis of lysozyme gene regulation led to the following conclusions:

1. Differential activity of several enhancer and silencer elements leads to the observed lysozyme gene activity in macrophages.
2. Within 250 bp of the chicken lysozyme RNA start site silencer, enhancer, glucocorticoid, and progesterone modules are tightly clustered and partially overlapping.
3. Silencer (data not shown), enhancer, and hormone responsive units can be composed of several modules.
4. The arrangement of modules can be changed although the precise stereo-specific alignment determines optimal function.
5. Synergistic interaction of transcription factors with steroid receptors leads to a pronounced steroid effect.

Our current interest is focused on interactions within these transcription factors and with other proteins.

References

Baniahmad, A., Muller, M., Steiner, Ch. & Renkawitz, R. (1987) EMBO J. 6:2297–2303
Barrera-Saldana, H., Takahashi, K., Vigneron, M., Wildeman, A., Davidson, I. & Chambon, P. (1985) EMBO J. 4:3839–3849
Bird, A.P. (1984) Nature (London) 307:503–504
Cato, A.C.B., Skroch, P., Weinmann, J., Butkeraitis, P. & Ponta, H. (1988) EMBO J. 7:1403–1408
Chodosh, L.A., Baldwin, A.S., Carthew, R.W. & Sharp, P.A. (1988) Cell 53:11–19
Danesch, U., Gloss, B., Schmid, W., Schütz, G., Schüle, R. & Renkawitz, R. (1987) EMBO J. 6:625–630
Hecht, A., Berkenstam, A., Stromstedt, P-E., Gustafsson, J.-A. & Sippel, A.E. (1988) EMBO J. 7:2063–2073
Lee, W., Haslinger, A., Karin, M. & Tjian, R. (1987) Nature (London) 325:368–372
Miksicek, R., Borgmeyer, U. & Nowock, J. (1987) EMBO J. 6:1355–1360
Payvar, F., DeFranco, D., Firestone, G.L., Edgar, B., Wrange, Ö., Okret, S., Gustafsson, J.-Å. & Yamamoto, K.R. (1983) Cell 356:381–392
Renkawitz, R., Schütz, G., von der Ahe, D. & Beato, M. (1984) Cell 37:503–510
Rosales, R., Vigneron, M., Macchi, M., Davidson, I., Xiao, J.H. & Chambon, P. (1987). EMBO J. 6:3015–3025
Scheidereit, C., Geisse, S., Westphal, H.M. & Beato, M. (1983) Nature (London) 304:749–752
Schüle, R., Muller, M., Otsuka-Murakami, H. & Renkawitz, R. (1988a) Nature (London) 332:87–90
Schüle, R., Muller, M., Kaltschmidt, C. & Renkawitz, R. (1988b) Science 242:1418–1420
Speck, N.A. & Baltimore, D. (1987) Mol. Cell. Biol. 7:1101–1110
Steiner, Ch., Muller, M., Baniahmad, A. & Renkawitz, R. (1987) Nucl. Acids Res. 15:4163–4177
Strähle, U., Klock, G. & Schütz, G. (1987) Proc. Natl. Acad. Sci. USA 84:7871–7875
Takahashi, K., Vigneron, M., Matthes, H., Wildeman, A., Zenke, M. & Chambon, P. (1986) Nature (London) 319:121–126
von der Ahe, D., Janich, S., Scheidereit, C., Renkawitz, R., Schütz, G. & Beato, M. (1985) Nature (London) 313:706–709

Chromatin Template Remodeling and Steroid Receptor Transactivation of MMTV

G.L. HAGER[1]

1 Introduction

The major current paradigm for our understanding of transcriptional regulation in eukaryotic cells is delineated within a set of concepts that are generally referred to as the protein-protein, or looping model. Under this view (Fig. 1), the proteins involved in gene activation, or repression, at a given locus are thought to interact cooperatively with each other in such a way that the stability of a given activating (or repressing) complex is initiated and maintained by direct contacts between the various members of the multifactorial group. The absence of one member of the complex would lower the affinity sufficiently to establish an equilibrium condition with the complex dissociated from the DNA regulatory sequence, and control of the target promoter would thus be exerted. According to this model, the primary function of DNA is simply to organize recognition sites for the various regulatory proteins in such a way that kinetically favorable protein-protein interactions can occur. That is, the DNA serves as a two-dimensional "string," assembling recognition sites for the appropriate factors in space so that the local concentration exceeds that required to facilitate assembly of the initiating complex.

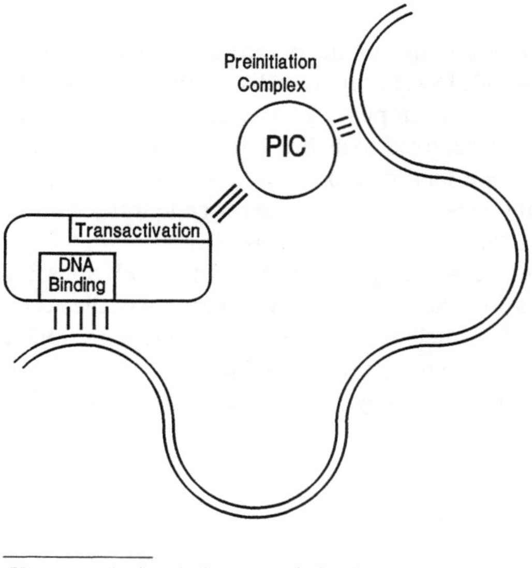

Fig. 1. Model for transactivation of transcription by protein-protein interactions at a distance (looping)

[1]Hormone Action & Oncogenesis Section, Laboratory of Experimental Carcinogenesis, National Cancer Institute, Bethesda, MD 20892, USA

40. Colloquium Mosbach 1989
Molecular Mechanisms of Hormone Action

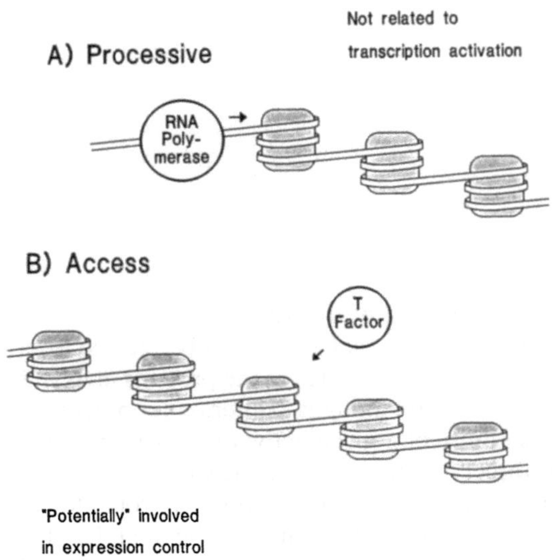

Fig. 2. Implications of the nucleosomal organization of DNA for processive events (replication and transcription), and for the access of regulatory factors

There is direct experimental evidence for this model in a few systems, primarily in prokaryotic systems (Ptashne 1986), and a great deal of indirect evidence, both in eukaryotic and prokaryotic prototypes. Several transcriptional regulatory proteins have now been molecularly cloned and subjected to mutational analysis. An emerging consensus suggests that many, if not all, transcriptionally active factors contain at least two domains, one responsible for the DNA-binding activity of the molecule, and a second specifying the transactivation function, presumably through protein-protein contacts (Fig. 1).

Although the role of DNA is limited under the looping paradigm to the positioning of factor binding sites, the organization of DNA in higher cells is quite complex. All DNA sequences in these cells are wrapped on a repeating structure, the nucleosome, resulting in an extended array commonly referred to as the 100 nm fiber. While this nucleoprotein structure can be further condensed into higher-order structures, it is thought that the single polynucleosomal array is the template for gene transcription. This structure is therefore often called "active chromatin" (Fig. 2).

The organization of DNA into nucleosomal particles could have obvious implications for the enzymology of replication and transcription. In particular, the access of factors to DNA sequences preformed in nucleosome structures could be inhibited. Greater interest was focused on this possible sequestration of DNA when it was clearly demonstrated that nucleosomes can be specifically positioned over certain DNA sequences (Simpson and Stafford 1983).

2 Phasing in the MMTV LTR

The classical technique of indirect end-labeling, coupled with the use of intranucleosomal-specific cleavage reagents permits one to examine the positioning of nucleosomes over a given region of DNA. Employing this methodology, we found

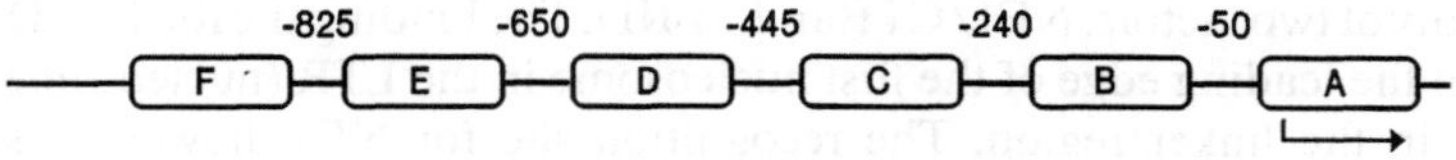

Fig. 3. Site-specific positioning of nucleosomes over the MMTV LTR

(Richard-Foy and Hager 1987) that nucleosomes are specifically positioned, or phased, over the LTR of mouse mammary tumor virus (MMTV) (Fig. 3). This retroviral LTR harbors a promoter that is responsive to stimulation by steroid hormones, acting through specific binding sites (HREs) for their cognate receptors found upstream of the promoter. Phasing over the MMTV LTR has now been found to be an intrinsic property of LTR DNA. Irrespective of the method by which DNA is inserted in the cell, specific positioning always occurs over the LTR, if the wild-type DNA sequence is maintained.

Nucleosome positioning is of particular interest in this case because the promoter can be rapidly activated. It has been known for some time that many so-called hypersensitive sites that appear in chromatin during cellular development correspond to nucleosome-free regions of DNA. It has been inferred that occupancy of a given DNA site by specific DNA-binding proteins is inconsistent with the formation of a nucleosome over the same region. Because these hypersensitive regions usually appear in concert with cell division and DNA replication, it has often been suggested that factors could effectively compete for a given site during periods of nucleosome instability, such as the passage of a DNA replication fork. Because MMTV transcription is induced so rapidly, the presence of positioned nucleosomes offered a new possibility for addressing this general issue.

3 Nucleosome Displacement

We have now discovered (Archer et al. submitted) that one nucleosome in this phased array is, in fact, specifically displaced during transcriptional activation by the activated glucocorticoid receptor. Furthermore, we have demonstrated that the displacement process takes place in the absence of DNA replication. This is the first documented example of a known regulatory factor, the glucocorticoid receptor, binding to its recognition sequence when that site is positioned on the surface of a nucleosome, and resulting in the active displacement of the octamer core from the DNA.

4 Factor Access to Nucleosomal DNA

These observations open several important questions to further analysis. Is the DNA on the surface of a nucleosome in general quite accessible to site-specific binding proteins, or can access be restricted in some cases? We now know that the accessibility enjoyed by the steroid receptor is not a general property of nucleosomally organized sequences.

When the MMTV promoter is activated, a transcriptional preinitiation complex is recruited to the TATA-Cap site region (Cordingley et al. 1987). This complex is composed minimally of two factors, NF1/CTF and TFIID. The binding site for TFIID is located partly on the leading edge of the first nucleosome in the LTR (nucleosome A), but primarily in the linker region. The recognition site for NF1, however, is positioned directly over the core of nucleosome B. The sites for both receptor and NF1, therefore, are localized to nucleosome B.

We have now found (Archer et al. submitted) that NF1 is unable to recognize its site when the DNA is organized into specifically positioned nucleosomes in vitro. This observation is in contrast to the findings of Perlmann and Wrange (1988), who found that glucocorticoid receptor could bind and footprint to its recognition site when the DNA was accurately positioned in a monosome. This latter observation is clearly consistent with the fact that receptor binding and transcriptional activation occur in vivo despite the presence of the phased nucleosomes.

5 Remodeling of MMTV LTR Chromatin During Activation

Given the several observations discussed above, we propose the following model regarding the interaction of glucocorticoid receptor and the preinitiation complex with the MMTV promoter when organized on phased nucleosomes (Fig. 4). The receptor is able to recognize and interact with its binding site on the surface of nucleosome B.

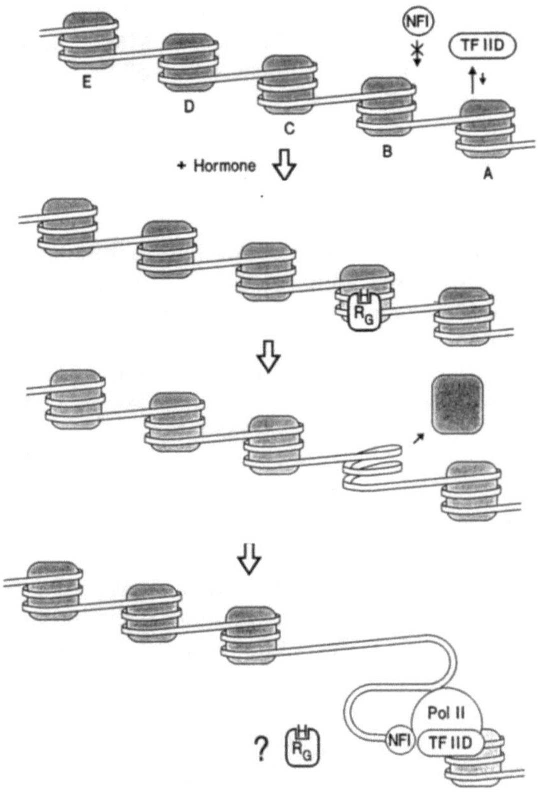

Fig. 4. Model for remodeling of MMTV LTR chromatin structure during transcriptional activation

This binding event initiates a process that results in displacement of that nucleosome. With available data, we cannot distinguish between direct displacement by the receptor, as opposed to displacement by a secondary process initiated by the receptor. In noninduced chromatin, factor NF1 is unable to access its recognition site. This partially explains the paradox that NF1 is completely excluded from LTR chromatin, even though it is present in the nucleus at a reasonably high concentration and in an active DNA binding form (Cordingley and Hager 1988). Indeed, the apparent affinity of NF1 for MMTV promoter DNA in vitro is much higher than that for glucocorticoid receptor. The binding site for transcription factor TFIID is probably available in chromatin, but the intrinsic affinity of this factor for the MMTV promoter is too weak for it to occupy its site alone.

Thus, the displacement of nucleosome B is necessary not to permit the binding of glucocorticoid receptor, but rather to allow the formation of the NF1-TFIID preinitiation complex. The mechanism by which the nucleosome is actually displaced remains an enigma at this time. Two general classes of mechanisms can be entertained. In one class, the receptor would have an intrinsic displacement activity, or be able to interact with molecules having such activity. Under this model, displacement could actually take place as a process separate and distinct from factor recruitment to the preinitiation complex, although this is not necessarily a possibility we favor. In a second class, one could propose that recruitment of factors into the NF1-TFIID complex by classic protein-protein interactions, either between the receptor and the complex itself or an intermediate communicating molecule, is the process that drives the B-region octamer core out of its position. Irrespective of which mechanism occurs (or some other as yet unspecified process), the important conclusion is that MMTV LTR chromatin is not a neutral structure with regard to factor access. The preinitiation complex is clearly excluded from in vivo nucleoprotein, and can be assembled on the promoter only by displacement of one nucleosome from the phased array.

It should be clearly stated that nothing in this model excludes the potential, or even probable, participation of classic protein-protein interactions between the receptor and some target component of the initiation cascade in the overall promoter activation process. What is distinct in this mechanism is the concept that the chromatin-organized promoter is a repressed structure, and active remodeling of chromatin is necessary in order to assemble the active initiation complex.

6 Mechanism of Factor Exclusion

Glucocorticoid receptor can recognize its site, but NF1/CTF is excluded when DNA is specifically positioned on an octamer core. What is the basis for this differential access? A very simple but intriguing mechanism is suggested by the relative position of the two recognition sites on the nucleosome (Fig. 5). The binding site for receptor is positioned over the core in such a manner that the side of the double helix which is addressed by the receptor as it binds is facing out of the histone octamer core. Conversely, the side of the helix addressed by NF1 is in fact facing directly into the nucleosome B core. Thus, one can propose that simple steric hindrance of access to the binding face is presented by the histone core in the case of NF1, while the initial contact sites for receptor are available and facing out of the core. This model is undoubtedly

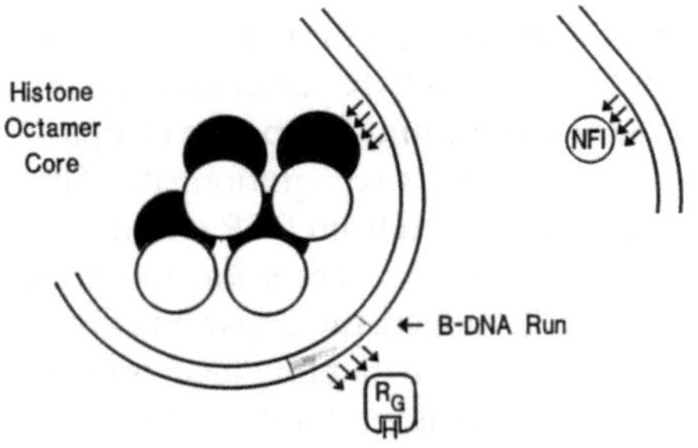

Fig. 5. Steric hindrance as a possible mechanism of differential factor access to nucleosomal DNA

oversimplified, since the path of DNA over the core is certainly not a simple curve of B-form DNA. Other factors, such as severe distortion of the helix at certain points on the surface of the nucleosome, must eventually be taken into consideration. The steric hindrance model is a useful beginning point, however, for mutational analysis of the factors involved in selective factor access. Given that we can now position octamer cores over their correct positions in vitro, and assay the accessibility of factors to DNA organized in such structures, means that the tools are now available to address this very interesting and potentially important question.

References

Cordingley, M.G. & Hager, G.L. (1988) Binding of multiple factors to the MMTV promoter in crude and fractionated nuclear extracts. Nucleic. Acids Res. 16:609–628

Cordingley, M.G., Riegel, A.T. & Hager, G.L. (1987) Steroid-dependent interaction of transcription factors with the inducible promoter of mouse mammary tumor virus in vivo. Cell 48:261–270

Perlmann, T. & Wrange, O. (1988) Specific glucocorticoid receptor binding to DNA reconstituted in a nucleosome. EMBO J. 7:3073–3079

Ptashne, M. (1986) Gene regulation by proteins acting nearby and at a distance. Nature (London) 322(6081):697–701

Richard-Foy, H. & Hager, G.L. (1987) Sequence specific positioning of nucleosomes over the steroid-inducible MMTV promoter. EMBO J. 6:2321–2328

Simpson, R.T. & Stafford, D.W. (1983) Structural features of a phased nucleosome core particle. Proc. Natl. Acad. Sci. USA 80:51–55

Hormone-Dependent Transcriptional Activation by Thyroid Hormone Receptors: Functional Homology with Steroid Hormone Receptors*

S.M. HOLLENBERG[1], C.C. THOMPSON[1,2], and R.M. EVANS[1]

1 Introduction

Thyroid hormones, thyroxine (T_4) and triiodothyronine (T_3), play critical roles in growth and homeostasis, influencing the production of many key regulatory proteins (Samuels et al. 1988). The actions of thyroid hormones, like the steroid hormones, are mediated through intracellular receptor proteins (Oppenheimer 1983; Samuels 1983). Complementary DNAs encoding homologues of the oncogene v-erbA have been isolated and were found to bind thyroid hormones with affinities characteristic of the native thyroid hormone receptor (Weinberger et al. 1986; Sap et al. 1986). The deduced primary amino acid sequences of these thyroid hormone binding proteins showed conservation with the steroid hormone receptors, members of a superfamily of ligand-binding transcription factors (Evans 1988).

Extensive analysis of the structural and functional properties of the glucocorticoid receptor and other steroid hormone receptors has led to the identification of discrete functional domains in these molecules (Evans 1988; Green and Chambon 1988; Beato 1989). A centrally located, 66 to 68 amino acid region with nine conserved cysteine residues, thought to form two "zinc finger" structures, is essential for DNA binding and determines target gene specificity (Green and Chambon 1987; Giguére et al. 1987; Kumar et al. 1987). The carboxyl terminal region of the receptors is required for ligand binding and regulates transformation of the receptors to an active form. The function of the amino terminus is not clear, although in some cases it appears to determine the efficiency and specificity of transactivation (Kumar et al. 1987; Hollenberg et al. 1987; Tora et al. 1988). The thyroid and steroid hormone receptors show the greatest similarity in the DNA binding domain, with 47% amino acid identity. The ligand binding domains have low but measurable homology, while the amino termini show no conservation of size or amino acid composition (Evans 1988).

The structural similarity between the thyroid hormone and steroid hormone receptors, in conjunction with their hormone-dependent regulation of gene expression, suggests that they may be functionally related as well. Therefore, we have addressed the issues of whether these thyroid hormone binding proteins can act as transcription factors, and in addition, whether they have the same functional architecture as the steroid hormone receptors.

[1]Howard Hughes Medical Institute, The Salk Institute for Biological Studies, La Jolla, CA 92138, USA
[1,2]University of California at San Diego, Department of Biology, La Jolla, CA 92093, USA

**Abbreviations*: T_3, 3,5,3'-triiodothyronine; CAT, chloramphenicol acetyltransferase; rTRα, rat alpha thyroid hormone receptor; hTRβ, human beta thyroid hormone receptor; hGR, human glucocorticoid receptor.

40. Colloquium Mosbach 1989
Molecular Mechanisms of Hormone Action

To analyze thyroid hormone receptor action, we have used a cotransfection assay to demonstrate that the cloned thyroid hormone receptors are capable of activating transcription from a thyroid hormone responsive promoter in a hormone-dependent manner. Furthermore, we have compared the functional properties of thyroid hormone and glucocorticoid receptors by exchanging domains between the human glucocorticoid and thyroid hormone receptors to produce functional hybrids. These experiments identify the domains of the thyroid hormone receptors responsible for promoter recognition and ligand binding, and demonstrate that the mechanism of the "hormone switch" for receptor activation is maintained between distantly related members of the steroid hormone receptor superfamily.

2 Results

2.1 *Trans-Activation by Thyroid Hormone Receptors*

To study transcriptional activation by thyroid hormone receptors, we have adapted the cotransfection assay originally devised to study the function of the human glucocorticoid receptor (hGR) (Giguére et al. 1986). In this assay, two plasmids are transfected into a receptor-deficient cell line, the first to express a receptor protein, the other to monitor transcription from a hormone responsive promoter. For the thyroid hormone receptor assay, the expression plasmid consists of the Rous sarcoma virus long terminal repeat (RSV-LTR) directing the expression of a cDNA encoding a thyroid hormone receptor (Fig. 1a). For the hGR, the reporter plasmid is the mouse mammary tumor virus long terminal repeat (MTV-LTR) fused to the bacterial chloramphenicol acetyltransferase (CAT) gene. To convert MTV-CAT to a thyroid hormone responsive reporter, an oligonucleotide containing a thyroid hormone response element (TRE) was inserted at position −191 of the MTV-LTR to generate TRE-CAT (Fig. 1a). This sequence, −169 to −200 of the rat growth hormone gene, specifically binds thyroid hormone receptors and can confer T_3 responsiveness to a heterologous promoter (Glass et al. 1987). Expression and reporter plasmids were cotransfected into CV-1 cells and CAT activity was measured in the absence and presence of T_3. Figure 1b shows that neither thyroid hormone receptor activates transcription from MTV-CAT. However, the presence of a TRE results in a thyroid hormone responsive MTV promoter. Induction of CAT activity is dependent on the cotransfection of a functional thyroid hormone receptor and the addition of T_3. In the presence of T_3, rTRα induces CAT activity approximately 15-fold, while hTRβ induces activity by about 5-fold.

2.2 *Chimeric Thyroid Hormone/Glucocorticoid Receptors*

A domain structure for the thyroid hormone receptors has been assigned based on homology with the steroid hormone receptors (Weinberger et al. 1986; Thompson et al. 1987), but the functions of these putative domains have not been directly demonstrated. We have constructed hybrid thyroid hormone/glucocorticoid receptors to compare the functional properties of the thyroid hormone and glucocorticoid receptors. To construct hybrid receptors, unique sites for the restriction enzymes Not I and Xho I were introduced to the amino and carboxyl terminal sides of the DNA binding domains of the hGR and hTRβ. These receptors, termed hGR$_{NX}$ (Giguére et

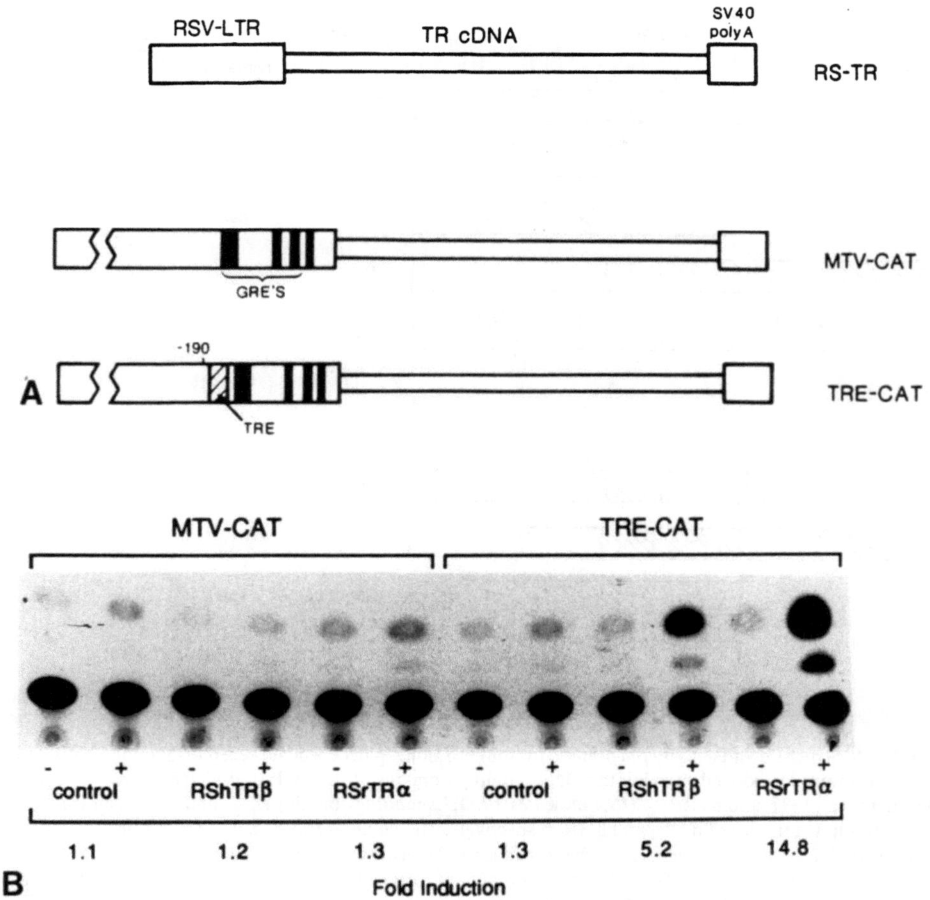

Fig. 1A,B. Trans-activation by thyroid hormone receptors. **A** Structure of thyroid hormone receptor expression and reporter plasmids. RS-TR has the RSV-LTR directing the expression of either the rTRα (Thompson et al. 1987) or hTRβ (Weinberger et al. 1986) cDNA followed by an SV40 polyadenylation signal. MTV-CAT, which has the MTV-LTR directing expression of the CAT gene, is shown with its glucocorticoid response elements (GREs) indicated as *black boxes* (Buetti and Kühnel 1986). TRE-CAT is identical to MTV-CAT except that it has a thyroid hormone response element from the rat growth hormone promoter (Glass et al. 1987) inserted at position –191. **B** Induction of CAT activity by thyroid hormone receptors. CV-1 cells were cotransfected with TRE-CAT and either rTRα or hTRβ expression plasmids. An expression vector with the rTRα cDNA in the antisense orientation was cotransfected as a negative control. Cultures were maintained in the absence (–) or presence (+) of 10^{-7} M T_3 and assayed for CAT activity. Fold induction, which is CAT activity in the presence of T_3 divided by CAT activity in the absence of T_3, is the average of five experiments

al. 1987) and hTRβ_{NX}, were used to create hybrids with all possible combinations of amino termini, DNA-binding domains, and ligand-binding domains (Fig. 2). The hybrid and parental receptors were assayed using both thyroid hormone and glucocorticoid responsive promoters, and treatment with no hormone, T_3, or the synthetic glucocorticoid dexamethasone. To assess the properties of the DNA binding domains, promoters that were selectively hormone responsive and of similar structure

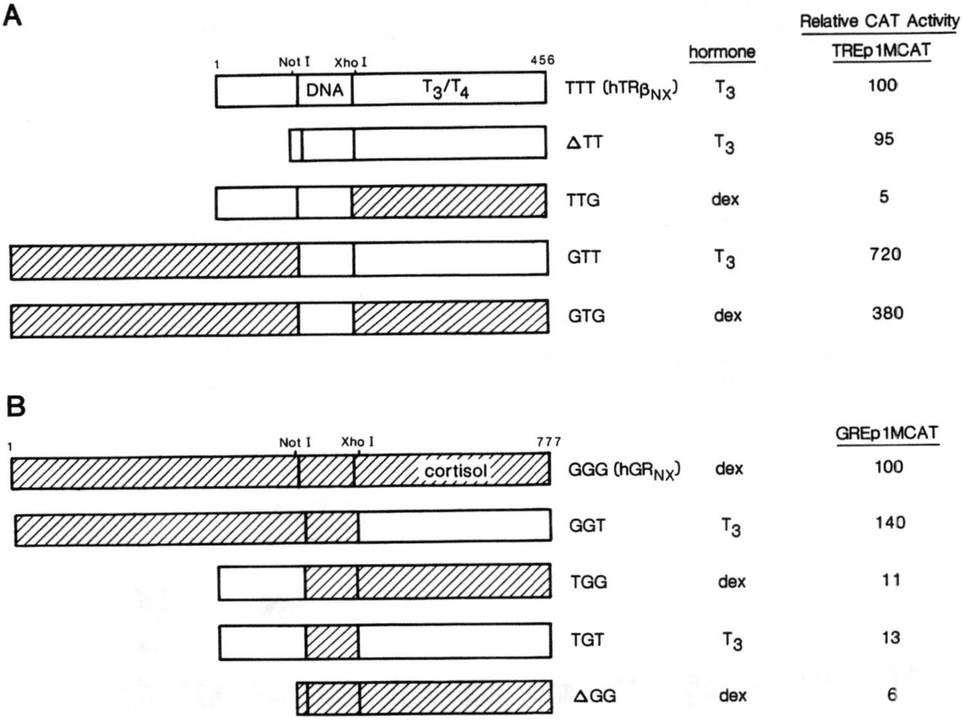

Fig. 2A,B. Structure and activity of chimeric thyroid hormone/glucocorticoid receptors. Unique NotI and XhoI sites were introduced flanking the DNA binding domains of the hGR and hTRβ to create hGR$_{NX}$ (Giguére et al. 1987) and hTRβ_{NX}. *DNA* indicates the DNA binding domain; T_3/T_4 and *cortisol* indicate the ligand binding domains of hTRβ and hGR, respectively. The *numbers above the boxes* indicate amino acid positions (Weinberger et al. 1986; Hollenberg et al. 1987). Hybrids are named by *letters* referring to the origin of the domain; for example, "T-G-T" has the amino and carboxyl termini of hTRβ (T-,-T) and the DNA binding domain of the hGR (-G-). All receptors were assayed on TREp1 MCAT and GREp1 MCAT in the absence and presence of 10^{-7} M T_3 and dexamethasone (*dex*). Values are shown only for receptor/reporter/ligand combinations that gave activation above background, and are the average of at least four experiments. Relative CAT activity is defined as induced CAT activity minus uninduced CAT activity normalized to either hTRβ_{NX} or hGR$_{NX}$. Activation is normalized on identical promoters so that activity is measured relative to hTRβ_{NX} for hybrids with an hTRβ DNA binding domain (**A**), and relative to hGR$_{NX}$ for hybrids with an hGR DNA binding domain (**B**)

were created by replacing the wild-type GREs in MTV-CAT with either a palindromic TRE (Glass et al. 1987) (TREp1 MCAT) or a palindromic GRE (Strähle et al. 1987) (GREp1 MCAT). Using these reporters, hTRβ_{NX} gives about a 20-fold induction, and hGR$_{NX}$ gives about a 150-fold induction (unpublished observations).

The structures and activities of the hybrid receptors are shown in Fig. 2. The receptors are divided into three sections, so that hTRβ_{NX} is called TTT, and hGR$_{NX}$ is called GGG. Hybrids are named by letters referring to the origin of the domain; for example, "T-G-T" has the amino and carboxyl termini of hTRβ (T-,-T) and the DNA binding domain of the hGR (-G-). Hybrids with a putative hTRβ DNA binding domain (TTG, GTT, GTG) activated transcription only from TREp1 MCAT, while hybrids with an hGR DNA binding domain (GGT, TGG, TGT) activated transcription only from GREp1 MCAT. This demonstrates that this region of hTRβ is analogous to the hGR DNA binding domain and is responsible for promoter recognition. Hybrid receptors with an hTRβ carboxyl terminus were activated only by T_3, while those with an hGR carboxyl terminus were activated only by dexamethasone. This is consistent with the identification of the carboxyl terminus as the region of the receptor that is responsible for hormone binding and activation specificity, and implies that thyroid and steroid hormones regulate the activities of their receptors through a similar mechanism.

Comparing the activities of the various hybrid receptors provides insight into the regions of the receptors that are responsible for trans-activation. Removal of the amino terminus of either thyroid hormone receptor has little impact on the ability of these receptors to activate this promoter (Fig. 2, unpublished observations). This suggests that the thyroid hormone receptors do not have a trans-activation function in their amino terminus but may, like the glucocorticoid and estrogen receptors (Hollenberg and Evans 1988; Webster et al. 1988), have this function in their carboxyl terminus. The indication that the TR amino terminus lacks a trans-activation function that is active on the MTV promoter is supported by the observation that hybrids in which the hTRβ amino terminus replaces that of the hGR have reduced activity. For example, TGG has about 10% of the activity of GGG, or close to the same activity as a GR with the amino terminus completely removed (ΔGG).

In contrast, hTRβ hybrids with an hGR amino terminus show a significant increase in activity. For example, GTT has approximately seven fold greater activity than TTT. This suggests that the hGR amino terminus possesses a trans-activation function that is not present in the TR amino terminus. Previous studies have shown that a discrete region of the hGR amino terminus (τ_1) is responsible for at least part of the transcriptional activity of the hGR (Hollenberg and Evans 1988; Godowski et al. 1988). To see if this activity could be transferred to a thyroid hormone receptor, cDNA sequences encoding amino acids 77–262 of the hGR were inserted in frame after amino acid 21 of the rTRα cDNA in one or multiple copies (Fig. 3). When the resulting hybrid receptors were assayed for trans-activation, the presence of a single τ_1 domain enhanced activity by approximately 14-fold. Additional τ_1 domains also increased activity, with maximal activation observed in the presence of two τ_1 domains. These results show that the activity of the thyroid hormone receptor can be augmented by the addition of a trans-activation domain from a heterologous receptor.

3 Discussion

The action of thyroid hormones is mediated by intracellular receptors which are associated with chromatin and are believed to be sequence-specific DNA binding proteins (DeGroot and Torresani 1975; Spindler et al. 1975; Glass et al. 1987; Koenig et al. 1987). These receptors are the primary effectors of thyroid hormone response, presumably by modulating the expression of specific genes in target cells (Samuels 1983; Evans 1988). Others have recently shown that the cloned thyroid hormone receptor subtypes can activate transcription from thyroid hormone responsive promoters (Koenig et al. 1985; Izumo and Mahdavi 1988; Forman et al. 1988). The

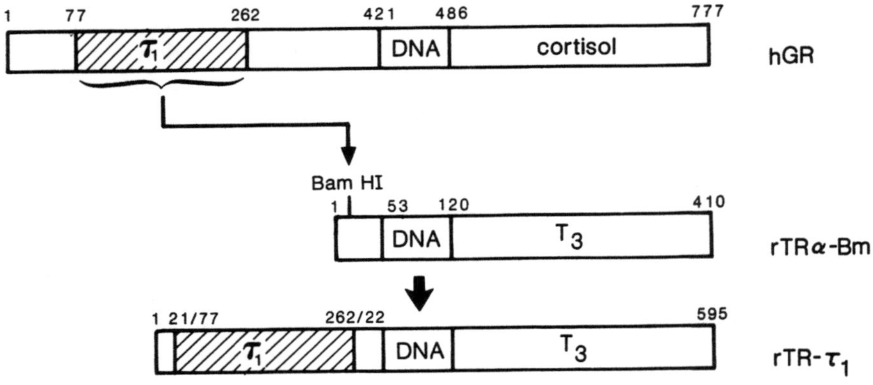

Receptor	# of τ_1 domains	Relative CAT Activity
hGR	1	0
rTRα-Bm	0	100
rTR-τ_1-1	1	1400
rTR-τ_1-2	2	4900
rTR-τ_1-3	3	2600
rTR-τ_1-4	4	1600

Fig. 3. Addition of τ_1 increases trans-activation by rTRα. A segment of the human glucocorticoid receptor cDNA encoding amino acids 77–262 (τ_1) (Hollenberg and Evans 1988) was inserted in the rTRα cDNA in one or multiple copies, and the resulting hybrids were assayed using TREp1 MCAT. The activity of the BamHI insertion derivative is approximately 90% of parental rTRα and represents an induction of about 25-fold. Relative CAT activity is determined as in Fig. 2.

results of this study demonstrate that both the alpha and beta forms of the thyroid hormone receptor can stimulate transcription from a thyroid hormone responsive promoter and that these molecules function in a manner that is mechanistically similar to that of the steroid hormone receptors.

The putative functional domains of the thyroid hormone receptors have been inferred based on structural homology with the glucocorticoid and estrogen receptors (Weinberger et al. 1986; Thompson et al. 1987). By creating hybrid thyroid hormone/glucocorticoid receptors, we were able to experimentally confirm these assignments. Thus, like the steroid receptors, the cysteine-rich region of the thyroid hormone receptors is also a DNA binding domain, since a hybrid glucocorticoid receptor harboring this domain activates transcription from a thyroid hormone responsive promoter. Previous studies had indicated that the carboxyl terminus was required for hormone binding (Muñoz et al. 1988), and consistent with this result, in hybrid receptors this region is sufficient to transfer the specificity of hormone activation. The observation that the ligand binding domain is responsible for the inducibility of the thyroid hormone receptors, and can be interchanged with that of a steroid hormone receptor, points out a striking conservation of function between these receptors. Since thyroid and steroid hormones are neither structurally nor biosynthetically related, it is not surprising that their ligand binding regions are only distantly related. However, the degree of structural conservation present in these hormone binding domains is sufficient to maintain hormone regulation of hybrid receptors.

Comparison of the transcriptional activities of the chimeric receptors indicates some differences as well as similarities between the thyroid hormone and glucocorticoid receptors. Like the glucocorticoid and estrogen receptors (Hollenberg and Evans 1988; Webster et al. 1988), the carboxyl terminus of the thyroid hormone receptors appears to play a role in trans-activation. However, the amino terminus of the glucocorticoid receptor contains an additional activation function (Hollenberg and Evans 1988; Godowski et al. 1988), which can be transferred to a thyroid hormone receptor to produce a receptor with greater than wild-type activity. This suggests that τ_1 can function in the context of the thyroid hormone receptor in a manner analogous to its action within the glucocorticoid receptor, perhaps by its association with other transcription factors.

The absence of a detectable function in the thyroid hormone receptor amino terminus does not mean that this region has no role in trans-activation on other promoters or in different cell types. The amino terminus is important for the activation efficiency and promoter specificity of some steroid hormone receptors. For example, the A and B forms of the progesterone receptor, which differ only in the amino terminus, can both activate transcription of the MTV promoter but only the A form can activate the ovalbumin promoter (Tora et al. 1988). Similarly, an amino terminally deleted estrogen receptor can regulate a vitellogenin-thymidine kinase hybrid promoter with the same efficiency as the wild-type receptor, but activates the pS2 promoter only 10% as well (Kumar et al. 1987). It is interesting to note that the amino terminus is the only part of the thyroid hormone receptors in which they are not homologous to each other. Like the estrogen and progesterone receptors, the amino terminus of the thyroid hormone receptors may have a function in the specificity or efficiency of activation.

Analysis of thyroid hormone receptor-mediated trans-activation reveals striking similarities with steroid hormone receptor action. Based on our studies, structural domains for promoter recognition, hormone binding, and trans-activation have been conserved. The ability to interchange domains between thyroid and steroid hormone receptors to produce functional hybrids indicates that the mechanisms involved in hormone inducible gene regulation have been conserved in these molecules.

Acknowledgments. We thank Dr. Christopher Glass for advice on thyroid hormone response elements, and Drs. Vincent Giguére and Klaus Damm for helpful suggestions. This work was supported by grants from the Howard Hughes Medical Institute and the National Institutes of Health.

References

Beato, M. (1989) Gene regulation by steroid hormones. Cell 56:335–344

Benbrook, D. & Pfahl, M. (1987) A novel thyroid hormone receptor encoded by a cDNA clone from a human testis library. Science 238:788–791

Buetti, E. & Kühnel, B. (1986) Distinct sequence elements involved in the glucocorticoid regulation of the mouse mammary tumor virus promoter identified by linker scanning mutagenesis. J. Mol. Biol. 190:379–389

Dayton, A.I., Selden, J.R., Laws, G., Dorney, D.J., Finan, J., Tripputi, P., Emanuel, B.S., Rovera G., Nowell P.C., & Croce, C.M. (1984) A human c-*erbA* oncogene homologue is closely proximal to the chromosome 17 breakpoint in acute promyelocytic leukemia. Proc. Natl. Acad. Sci. USA 81:4495–4499

DeGroot, L.J. & Torresani, J. (1975) Triiodothyronine binding to isolated liver cell nuclei. Endocrinology 96:357–369

Evans, R.M. (1988) The steroid and thyroid hormone receptor superfamily. Science 240:889–895

Forman, B.M., Yang, C.-R., Stanley, F., Casanova, J., & Samuels, H.H. (1988) c-*erbA* protooncogenes mediate thyroid hormone-dependent and independent regulation of the rat growth hormone and prolactin genes. Mol. Endocrinol. 2:902–911

Giguére, V., Hollenberg, S.M., Rosenfeld, M.G., & Evans, R.M. (1986) Functional domains of the human glucocorticoid receptor. Cell 46:645–652

Giguére, V., Ong, E.S., Segui, P., & Evans, R.M. (1987) Identification of a receptor for the morphogen retinoic acid. Nature (London) 330:624–629

Glass, C.K., Franco, R. Weinberger, C., Albert, V.R., Evans, R.M., & Rosenfeld, M.G. (1987) A c-*erbA* binding site in rat growth hormone gene mediates *trans*-activation by thyroid hormone. Nature (London) 329:738–741

Godowski, P.J., Picard, D., & Yamamoto, K.R. (1988) Signal transduction and transcriptional regulation by glucocorticoid receptor-LexA fusion proteins. Science 241:812–816

Green, S. & Chambon, P. (1987) Oestradiol induction of a glucocorticoid response gene by a chimaeric receptor. Nature (London) 325:75–78

Green, S. & Chambon, P. (1988) Nuclear receptors enhance our understanding of transcription regulation. Trends Genet. 4:309–314

Hollenberg, S.M. & Evans, R.M. (1988) Multiple and cooperative *trans*-activation domains of the human glucocorticoid receptor. Cell 55:899–906

Hollenberg, S.M., Giguére, V., Segui, P., & Evans, R.M. (1987) Colocalization of DNA-binding and transcriptional activation functions in the human glucocorticoid receptor. Cell 49:39–46

Izumo, S. & Mahdavi, V. (1988) Thyroid hormone receptor α isoforms generated by alternative splicing differentially activate myosin HC gene transcription. Nature (London) 334:539–542

Koenig, R.J., Brent, G.A., Warne, R.L., Larsen P.R., & Moore, D.D. (1987) Thyroid hormone receptor binds to a site in the rat growth hormone promoter required for induction by thyroid hormone. Proc. Natl. Acad. Sci. USA 84:5670–5674

Koenig, R.J., Warne, R.L., Brent, G.A., Harney, J.W., Larsen, P.R., & Moore, D.D. (1988) Isolation of a cDNA clone encoding a biologically active thyroid hormone receptor. Proc. Natl. Acad. Sci. USA 85:5031–5035

Kumar, V., Green, S., Stack, G., Berry, M., Jin, J.-R., & Chambon, P. (1987) Functional domains of the human estrogen receptor. Cell 51:941–951

Lazar, M.A., Hodin, R.A., Darling, D.S., & Chin, W.W. (1988) Identification of a rat c-erbAα-related protein which binds deoxyribonucleic acid but does not bind thyroid hormone. Mol. Endocrinol. 2:893–901

Mitsuhashi, T., Tennyson, G.E., & Nikodem, V.M. (1988) Alternative splicing generates messages encoding rat c-erbA proteins that do not bind thyroid hormone. Proc. Natl. Acad. Sci. USA 85:5804–5808

Muñoz, A., Zenke, M., Gehring, U., Sap, J., Beug, H., & Vennström, B. (1988) Characterization of the hormone-binding domain of the chicken c-erbA/thyroid hormone receptor protein. EMBO J. 7:155–159

Nakai, A., Seino, S., Sakurai, A., Szilak, I., Bell, G.I., & DeGroot, L.J. (1988) Characterization of a thyroid hormone receptor expressed in human kidney and other tissues. Proc. Natl. Acad. Sci. USA 85:2781–2785

Oppenheimer, J.H. (1983) The nuclear receptor-triiodothyronine complex: relationship to thyroid hormone distribution, metabolism, and biological action. In: Oppenheimer, J.H. & Samuels, H.H. (eds) Molecular basis of thyroid hormone action. Academic Press, New York London, pp. 1–34

Oppenheimer, J.H., Schwartz, H.L., & Surks, M.I. (1974) Tissue differences in the concentration of triiodothyronine nuclear binding sites in the rat: liver, kidney, pituitary, heart, brain, spleen, testis. Endocrinology 95:897–903

Samuels, H.H. (1983) Identification and characterization of thyroid hormone receptors and action using cell culture techniques. In: Oppenheimer, J.H. & Samuels, H.H. (eds) Molecular basis of thyroid hormone action. Academic Press, New York London, pp. 35–64

Samuels, H.H., Forman, B.M., Horowitz, Z.D., Ye, Z.S. (1988) Regulation of gene expression by thyroid hormone. J. Clin, Invest. 81:957–967

Sap, J., Muñoz, A., Damm, K., Goldberg, Y., Ghysdael, J., Leutz, A., Beug, H., & Vennstrom, B. (1986) The c-*erbA* protein is a high-affinity receptor for thyroid hormone. Nature (London) 324:635–640

Spindler, B.J., MacLeod, K.M., Ring, J., & Baxter, J.D. (1975) Thyroid hormone receptors. J. Biol. Chem. 250:4113–4119

Spurr, N.K., Solomon, E., Jansson, M., Sheer, D., Goodfellow, P.N., Bodmer, W.F., & Vennstrom, B. (1984) Chromosomal localisation of the human homologues to the oncogenes *erb*A and B. EMBO J. 3:159–163

Strähle, U., Klock, G., & Schütz, G. (1987) A DNA sequence of 15 base pairs is sufficient to mediate both glucocorticoid and progesterone induction of gene expression. Proc. Natl. Acad. Sci. USA 84:7871–7875

Thompson, C.C., Weinberger, C., Lebo, R., & Evans, R.M. (1987) Identification of a novel thyroid hormone receptor expressed in the mammalian central nervous system. Science 237:1610–1614

Tora, L., Gronemeyer, H., Turcotte, B., Gaub, M.-P., & Chambon, P. (1988) The N-terminal region of the chicken progesterone receptor specifies target gene activation. Nature (London) 333:185–188

Webster, N.J.G., Green, S., Jin, J.R., & Chambon, P. (1988) The hormone-binding domains of the estrogen and glucocorticoid receptors contain an inducible transcription activation function. Cell 54:199–207

Weinberger, C., Hollenberg, S.M., Rosenfeld, M.G., & Evans, R.M. (1985) Domain structure of human glucocorticoid receptor and its relationship to the c-*erb-A* oncogene product. Nature (London) 318:670–672

Weinberger, C., Thompson, C.C., Ong, E.S., Lebo, R., Gruol, D.J., & Evans, R.M. (1986) The c-*erb-A* gene encodes a thyroid hormone receptor. Nature (London) 324:641–646

Subunit Structure of the Glucocorticoid Receptor

U. GEHRING[1], M. REXIN[1], W. BUSCH[1], B. SEGNITZ[1], and G. ZINK[1]

1 Introduction

The molecular weights of specific receptors for glucocorticoids and other steroid hormones depend heavily on the conditions of investigation. High molecular weight forms of 300 000 Da and above are detected in extracts of target cells under very mild conditions of analysis which avoid subunit dissociation and denaturation. Routinely, such receptor forms are analyzed by gel permeation chromatography and sedimentation in glycerol or sucrose gradients (Sherman and Stevens 1984). On the other hand, the hormone-binding polypeptides are most easily analyzed by affinity labeling with a radiolabeled steroid derivative and subsequent SDS gel electrophoresis (Gronemeyer 1988). These receptor polypeptides form a protein family, all members of which have a similar domain arrangement even though the sizes of the polypeptides vary considerably amongst the receptors of different hormone specificities (Gehring 1987; Evans 1988). They all contain a central DNA binding domain of less than 70 amino acid residues and a carboxy terminal region of more than 200 amino acids which is involved in hormone binding.

The high molecular weight structure of the wild-type glucocorticoid receptor has a size of 330 000 Da and contains only *one* steroid binding polypeptide of $M_r \sim 100\,000$ (Gehring and Arndt 1985; Okret et al. 1985; Gehring et al. 1987) which is associated with other protein subunits. One of the most striking features of the large receptor form is the fact that it is unable to interact with DNA.

2 Activation to DNA Binding

In order for DNA binding ability to be expressed the high molecular weight receptor-glucocorticoid complex needs to be activated. Activation is brought about by various treatments, most notably by warming the receptor solution or by increasing the ionic strength. The nonactivated wild-type receptor of glucocorticoid-sensitive S49.1 mouse lymphoma cells has a Stokes radius of about 80 Å (Table 1). In the experiment of Fig. 1 we submitted a partially activated receptor sample to gel filtration and, as expected, observed two peaks with Stokes radii of ~ 80 and ~ 60 Å, respectively. Prior filtration through DNA-cellulose completely removed the 60 Å peak but left the 80 Å material unadsorbed. This result unequivocally demonstrates the DNA binding ability of the 60 Å, but not of the 80 Å receptor form.

[1]Institut für Biologische Chemie der Universität Heidelberg, Im Neuenheimer Feld 501, 6900 Heidelberg, Fed. Rep. of Germany

40. Colloquium Mosbach 1989
Molecular Mechanisms of Hormone Action

Table 1. Cross-linking of the high molecular weight receptor

	Treatment[a]	Stokes radius[b] (Å)	Molecular weight[b]	DNA binding
Control	–	80.6*	328 000*	–
	–	59.6	116 000	+
Cross-linking in cell extracts	DSP	82.3	316 000	–
	DSP, cleavage	60.2	~ 120 000	+
	Cu^{2+}/oP	80.2	335 000	–
	Cu^{2+}/oP, cleavage	61.6	~ 125 000	+
	DEPC	80.9	~ 330 000	–
Cross-linking in intact cells	DSP	81.2	~ 320 000	
	DSP, cleavage	60.0	~ 120 000	

[a] DSP = dithiobis(succinimidyl propionate); Cu^{2+}/oP = cupric/o-phenanthroline; DEPC = diethyl pyrocarbonate.
[b] Analysis was in the presence of 300 mM KCl except in *.

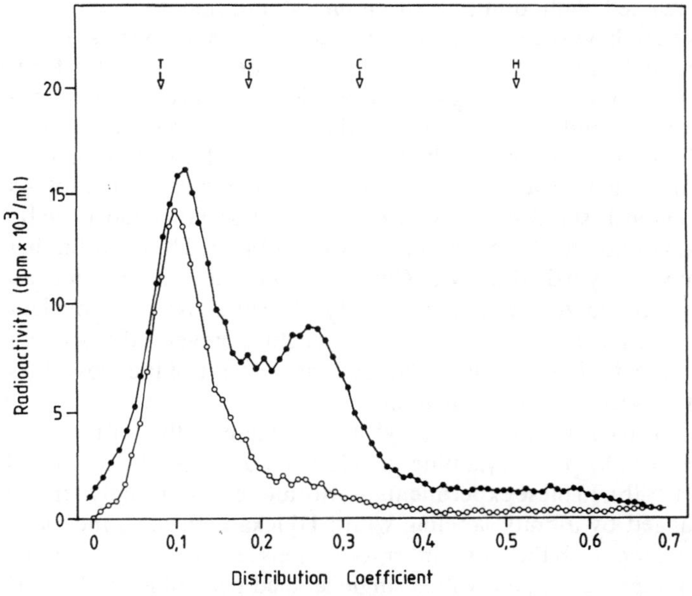

Fig. 1. Gel filtration on Sephacryl S-300 of the wild-type S49.1 receptor complex with [^{3}H]triamcinolone acetonide. Following a 30 min incubation at 20°C, identical samples were analyzed at low ionic strength in the cold either without further treatment (●) or after passing through a DNA-cellulose column (○). Markers for Stokes radii were thyroglobulin (*T*; 86.1 Å), *β*-galactosidase (*G*; 68.5 Å); catalase (*C*; 52.3 Å), and hemoglobin (*H*; 32.1 Å). (Data from Gehring et al. 1987)

Treatment with, for example, 300 mM KCl likewise results in conversion of the 80 Å wild-type receptor form to a species of 60 Å Stokes radius which is able to bind to DNA (Gehring et al. 1987). The Stokes radii and sedimentation coefficients were used to compute the molecular weights (Table 1). The data clearly show that activation to the DNA binding state occurs concomitantly with the dissociation of the 330 000 Da wild-type receptor form to a species of $M_r \sim 116\,000$.

3 Chemical Cross-Linking of High Molecular Weight Receptors

If indeed receptor dissociation and activation to DNA binding are tightly coupled events, then it should be possible to inhibit activation by preventing dissociation through chemical cross-linking of receptor subunits. This has been achieved by use of several reagents (Rexin et al. 1988a). To this end, bifunctional N-hydroxysuccinimidyl esters and bisimidates, both of which react with ε-amino groups of lysine residues, turned out to be particularly useful (Bäumert and Fasold 1989). In addition, sulfhydryl groups in protein subunits can be linked to each other by mild oxidation with atmospheric oxygen in conjunction with the complex of cupric ions with o-phenanthroline.

Figure 2 depicts an experiment in which the wild-type receptor complex of S49.1 lymphoma cells was reacted with dithiobis(succinimidyl propionate). Upon gel filtration at high ionic strength two peaks were observed corresponding to cross-linked (~ 80 Å) and non-cross-linked (~ 60 Å) material. Most of the 60 Å receptor form was again removed by prior filtration through DNA-cellulose but the cross-linked receptor did not bind to DNA and thus appeared in the effluent. Cross-linking with dithiobis(succinimidyl propionate) has the advantage that the molecule bridging the receptor subunits can subsequently be cleaved by reduction with mercaptoethanol. As shown in Fig. 2B the previously stabilized 80 Å receptor form was again dissociated by high salt to yield a Stokes radius of 60 Å. Similar results were obtained by cross-linking the larger receptor structure by oxidation with Cu^{2+}/o-phenanthroline (Rexin et al. 1988a). The data are summarized in Table 1. They clearly show that covalent cross-linking of the subunits within the large receptor structure prevents dissociation and hence activation to DNA binding ability. Subsequent cleavage of the cross-links again permits receptor dissociation and activation.

In an alternative approach we used dimethyl suberimidate as the bifunctional cross-linking reagent and SDS polyacrylamide gel electrophoresis as the analytical technique (Rexin et al. 1988b). In these experiments the hormone-binding polypeptide of the receptor was tagged by affinity labeling with [^{3}H]dexamethasone mesylate. Figure 3 shows that treatment with the cross-linker for various lengths of time resulted in progressive cross-linking. Starting out with a single labeled peak of $M_r \sim 100\,000$ (Fig. 3A) increasingly complex patterns were obtained which finally shifted toward a major species of ~ 350 000 Da (Fig. 3D). This corresponds to the fully cross-linked receptor material described above. A total of four intermediate forms were seen consistently, albeit at different intensities. Table 2 summarizes the molecular weight data and presents our interpretation. We conclude that the large receptor structure is a heterotetramer which contains one steroid-binding polypeptide of $M_r \sim 100\,000$ Da associated with two subunits of M_r 90 000 and one polypeptide of ~ 50 000 Da. Our

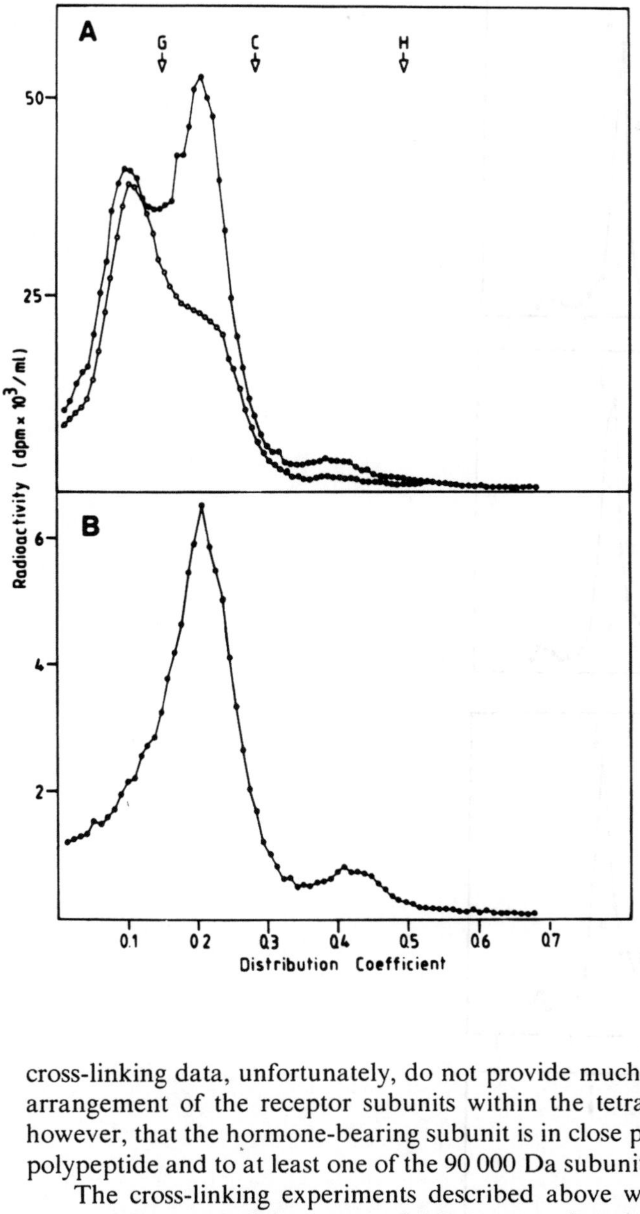

Fig. 2A,B. Gel filtration in the presence of 300 mM KCl of wild-type S49.1 receptor complexes with [^{3}H]triamcinolone acetonide following treatment with dithiobis (succinimidyl propionate. Identical samples were either applied directly to Sephacryl S-300 columns (A,●) or after passing through DNA-cellulose (A,○). In **B**, the cross-linked 80 Å material was isolated and submitted to reductivecleavagewithmercaptoethanol prior to gel filtration. (Data from Rexin et al. 1988a)

cross-linking data, unfortunately, do not provide much information on the relative arrangement of the receptor subunits within the tetrameric structure. It is clear, however, that the hormone-bearing subunit is in close proximity to the ~ 50 000 Da polypeptide and to at least one of the 90 000 Da subunits.

The cross-linking experiments described above were also carried out with a mutant glucocorticoid receptor of S49.1 mouse lymphoma cells in which the hormone-binding polypeptide is truncated; the amino terminal half of the molecule is missing but the DNA binding and hormone-binding domains are intact. This mutant receptor also occurs in a high molecular weight nonactivated form which behaves in cross-linking experiments identical to the wild-type (Rexin et al. 1988a,b). Our data with the mutant receptor substantiate the conclusion that the high molecular weight form of glucocorticoid receptors is a heterotetramer.

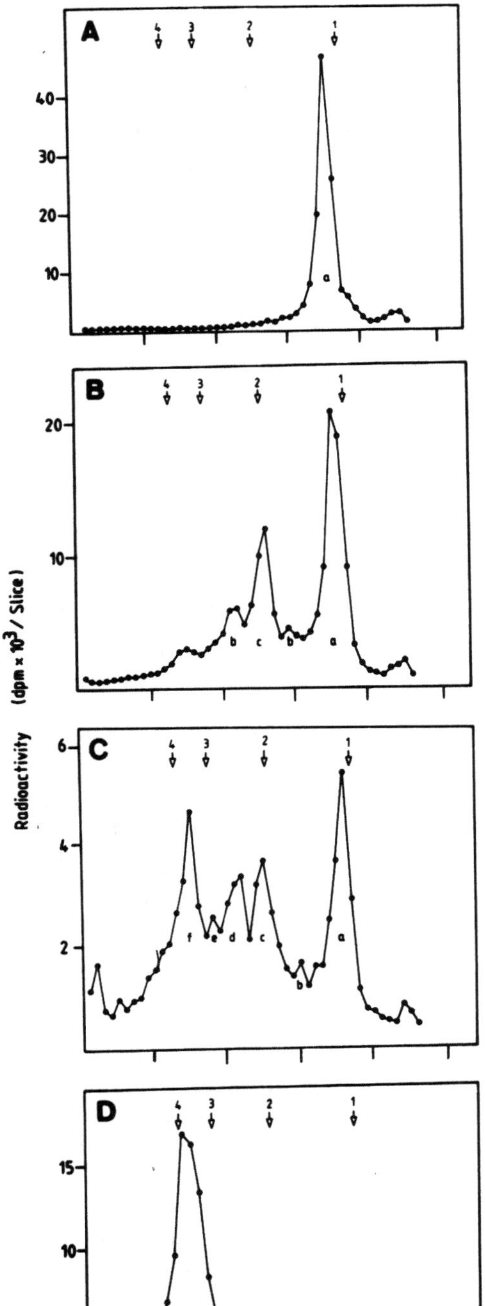

Fig. 3A-D. SDS polyacrylamide gel electrophoresis of wild-type S49.1 receptor complexes with [^{3}H]dexamethasone mesylate following treatment with dimethyl suberimidate in the cold for 0 (**A**), 20 (**B**), 40 (**C**), and 60 (**D**) min. Multimers of phosphorylase subunit are indicated by *arrows 1* to *4*. (Data from Rexin et al. 1988b)

Table 2. Receptor cross-linking with dimethyl suberimidate

	Molecular weight of labeled receptor species[a]	Subunit composition[b] (interpretation)
a)	104 000 ± 5 000	R
b)	149 000 ± 5 000	R + 50 kDa
c)	194 000 ± 10 000	R + 90 kDa
d)	241 000 ± 9 000	R + 90 kDa + 50 kDa
e)	301 000 ± 12 000	R + 90 kDa + 90 kDa
f)	349 000 ± 8 000	R + 90 kDa + 90 kDa + 50 kDa

[a] Letters a to f pertain to the labeled peaks of Fig. 3.
[b] R refers to the wild-type steroid-binding polypeptide of ~ 100 000 Da.

The heat shock protein of 90 000 (hsp90) has been detected in high molecular weight steroid receptors of different hormone specificities (Pratt 1987; Toft et al. 1987). We produced a polyclonal antibody directed against hsp90 from mouse lymphoma cells (Rexin et al. 1988a). By use of this antiserum we were able to identify in immunoblots hsp90 as one of the components of the high molecular weight glucocorticoid receptor which was cross-linked by treatment with dithiobis (succinimidyl propionate) or by oxidation with Cu^{2+}/o-phenanthroline. We therefore assume that the two subunits of 90 000 Da contained in the receptor heterotetramer are in fact two molecules of hsp90. This is consistent with data of Mendel and Orti (1988) who reported a roughly 1:2 ratio of steroid-binding subunit to hsp90 in WEHI-7 mouse cells, a cell line which is related to the S49.1 cells which we used in our studies.

As to the identity of the receptor subunit of ~ 50 000 Da there is as yet no detailed information at hand. Since the cross-linker dimethyl suberimidate is known to have a high degree of selectivity for amino groups in proteins we infer that this subunit is of polypeptide nature. It appears likely that the ~ 50 000 Da subunit detected in our cross-linking experiments is identical to a 59 000 Da protein which had been previously described as a component common to the high molecular weight forms of several steroid receptor types (Tai et al. 1986).

4 Is RNA a Component of the High Molecular Weight Receptor?

The involvement of RNA of the approximate size of tRNA in the high molecular weight glucocorticoid receptor structure has been suggested by several groups, in particular in receptors of liver cells (Economidis and Rousseau 1985; Webb and Litwack 1986; Sablonnière et al. 1988; Unger et al. 1988). In order to test this possibility in our S49.1 lymphoma cell system we used UV irradiation for inducing cross-links. In order to prevent RNA degradation by ubiquitous RNases we included in these experiments diethyl pyrocarbonate. We indeed observed cross-linking and stabilization of the ~80 Å wild-type receptor form (Rexin et al. 1988a). Careful control experiments, however, revealed that cross-linking was not caused by UV light but rather by exposure to diethyl pyrocarbonate (Table 1). This reagent apparently produced isopeptide linkages between receptor subunits.

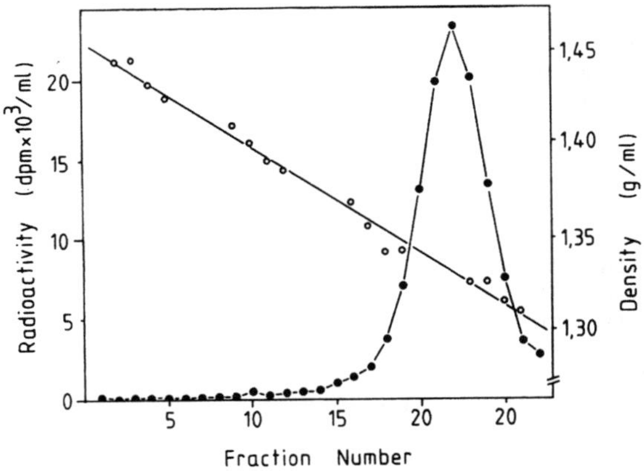

Fig. 4. Isopycnic density gradient centrifugation of cross-linked wild-type S49.1 receptor complexes with [^{3}H]dexamethasone mesylate. The receptor complex was dialyzed against 2% formaldehyde and submitted to gel permeation chromatography on Sephacryl S-300. The 80 Å peak was collected and the material mixed with CsCl to give a density of approximately 1.38 g ml^{-1}. Isopycnic centrifugation was at 65 000 rpm for 18 h at 15 °C in a Beckman VTi65 rotor; 190 μl fractions were collected and assayed for density and radioactivity. Control experiments with delipidated bovine serum albumin gave a density of 1.31 g ml^{-1}

In a more rigorous test for participation of RNA we used cross-linking with formaldehyde. In the experiment of Fig. 4 the high molecular weight receptor complex was labeled with [^{3}H]dexamethasone mesylate, treated with formaldehyde and subjected to analysis in an isopycnic CsCl gradient. The receptor material consistently banded at a density of 1.33 g ml^{-1}. We obtained the same density after prior treatment of the 80 Å receptor complex with ribonuclease A or after activation to the DNA binding state. We therefore conclude that the wild-type glucocorticoid receptor of S49.1 mouse lymphoma cells does not contain RNA to any significant amount. This result is in contrast to several reports dealing with glucocorticoid receptors of liver cells (Economidis and Rousseau 1985; Sablonnière et al. 1988; Unger et al. 1988) and thus raises the question whether receptors of lymphatic and hepatic cells may have different subunit structure. This does not appear likely. We rather assume that this discrepancy is due to subtle differences in techniques or to unspecific association of RNA to the receptor in liver cell extracts. In fact, association of the activated glucocorticoid receptor of pituitary cells with several tRNA species has been observed (Ali and Vedeckis 1987).

5 High Molecular Weight Receptor Form in Intact Cells

We used chemical cross-linking techniques to determine whether or not the high molecular weight glucocorticoid receptor form does preexist in intact cells. Following treatment of S49.1 cells with radiolabeled hormone in the cold we reacted the cells with dithiobis(succinimidyl propionate) in the presence of 5% DMSO. After removal of excess reagent the cells were ruptured and the extract analyzed by gel permeation chromatography in the presence of 300 mM KCl (Rexin et al. 1988a). We observed the stabilized 80 Å receptor form which was indistinguishable by biochemical and immunochemical techniques from that obtained by receptor cross-linking in cell extracts (Table 1). Similar results were also obtained when we treated whole cells with dimethyl suberimidate. Even though cross-linking was less efficient than in cell extracts we observed in SDS gels some of the intermediate receptor forms as in Fig. 3, in addition to the fully cross-linked heterotetramer of ~ 350 000 Da (Rexin et al. 1988b).

Taken together these data clearly show that the high molecular weight receptor of ~ 330 000 Da is not generated artifactually upon cell homogenization but rather preexists within intact cells. This supports the view that activation to the DNA binding state occurs in whole cells under physiological conditions (Munck and Foley 1979) and that this involves dissociation of the large receptor structure. The physiological significance of the nonhormone-binding subunits is still somewhat uncertain. It appears likely that they keep the receptor polypeptide in a configuration in which it is unable to translocate to the nucleus and thus does not interact with chromatin as long as the hormone is absent. Upon hormone binding to the large receptor structure activation occurs; this releases the steroid-bearing subunit for chromatin interaction and transcriptional regulation.

References

Ali, M. & Vedeckis, W.V. (1987) Interaction of RNA with transformed glucocorticoid receptor. J. Biol. Chem. 262:6771–6784

Bäumert, H.G. & Fasold, H. (1989) Cross-linking techniques. Meth. Enzymol. 172:584–609

Economidis, I.V. & Rousseau, G.G. (1985) Association of the glucocorticoid hormone receptor with ribonucleic acid. FEBS Lett. 181:47–52

Evans, R.M. (1988) The steroid and thyroid hormone receptor superfamily. Science 240:889–895

Gehring, U. (1987) Steroid hormone receptors: biochemistry, genetics and molecular biology. Trends Biochem. Sci. 12:399–402

Gehring, U. & Arndt, H. (1985) Heteromeric nature of glucocorticoid hormone receptors. FEBS Lett. 179:138–142

Gehring, U., Mugele, K., Arndt, H. & Busch, W. (1987) Subunit dissociation and activation of wild-type and mutant glucocorticoid receptors. Mol. Cell Endocrinol. 53:33–44

Gronemeyer, H. (1988) Affinity labelling and cloning of steroid and thyroid hormone receptors. Horwood, Chichester, UK

Mendel, D.B. & Orti, E. (1988) Isoform composition and stoichiometry of the ~ 90 kDa heat shock protein associated with glucocorticoid receptors. J. Biol. Chem. 263:6695–6702

Munck, A. & Foley, R. (1979) Activation of steroid hormone-receptor complexes in intact target cells in physiological conditions. Nature (London) 278:752–754

Okret, S., Wikström, A.-C. & Gustafsson, J.-Å. (1985) Molybdate-stabilized glucocorticoid receptor: evidence for a receptor heteromer. Biochemistry 24:6581–6586

Pratt, W.B. (1987) Transformation of glucocorticoid and progesterone receptors to the DNA-binding state. J. Cell Biochem. 35:51–68

Rexin, M., Busch, W. & Gehring, U. (1988a) Chemical cross-linking of heteromeric glucocorticoid receptors. Biochemistry 27:5593–5601

Rexin, M., Busch, W., Segnitz, B. & Gehring, U. (1988b) Tetrameric structure of the nonactivated glucocorticoid receptor in cell extracts and intact cells. FEBS Lett. 241:234–238

Sablonnière, B., Economidis, I.V., Lefebvre, P., Place, M., Richard, C., Formstecher, P., Rousseau, G.G. & Dautrevaux, M. (1988) RNA binding to the untransformed glucocorticoid receptor. Eur. J. Biochem. 177:371–382

Sherman, M.R. & Stevens, J. (1984) Structure of mammalian steroid receptors: evolving concepts and methodological developments. Annu. Rev. Physiol. 46:83–105

Tai, P.-K.K., Maeda, Y., Nakao, K., Wakim, N.G., Duhring, J.L. & Faber L.E. (1986) A 59-kilodalton protein associated with progestin, estrogen, androgen, and glucocorticoid receptors. Biochemistry 25:5269–5275

Toft, D.O., Sullivan, W.P., Mc Cormick, D.J. & Riehl, R.M. (1987) Heat shock proteins and steroid hormone receptors. Biochem. Actions Hormone 14:293–316

Unger, A.L., Uppaluri, R., Ahern, S., Colby, J.L. & Tymoczko, J.L. (1988) Isolation of ribonucleic acid from the unactivated rat liver glucocorticoid receptor. Mol. Endocrinol. 2:952–958

Webb, M.L. & Litwack, G. (1986) Association of RNA with the glucocorticoid receptor and possible role in activation. Biochem. Actions Hormone 13:379–402

Modulation of Glucocorticoid Hormone Action by Oncogenes and Peptide Hormones

B. GRONER[1], W. HÖCK[1], W. DOPPLER[1], and R. BALL[1]

1 Introduction

Studies on the mode of action of steroid hormones have had a pioneering role in the elucidation of eukaryotic gene regulation. Several principles, established by molecular endocrinologists, have proven to be of general validity, and have advanced our perception of transcriptional control. Two aspects have received major attention: (1) The identification of DNA sequence elements which confer hormonal inducibility to adjacent promoters. These hormone response elements (HRE) have been defined as specific binding sites for receptor molecules (Beato 1989). (2) The structure of the steroid receptor molecules has been elucidated in molecular detail (Evans 1988; Green and Chambon 1988). Gene cloning and transfer experiments have given us a precise picture of the domain structure of these molecules and of their interactions with hormone response elements. The concentration on these molecular aspects has increased the resolution of our understanding, but leaves a broad field of the dynamics and physiology of hormone action to be uncovered. This chapter will focus on the fate of the glucocorticoid receptor upon hormonal activation, and the regulation of receptor concentration and receptor phosphorylation will be described.

2 Results and Discussion

2.1 Glucocorticoid Receptor Modification and Concentration Is Regulated by Ligand Interaction

The mouse glucocorticoid receptor is an inducible transcription factor of 100 kDa which resides in the cytoplasm of many cell types (Danielson et al. 1986). Upon ligand association it is translocated to the nucleus where it assumes its role in the process of transcriptional regulation. The GR is a member of a large gene family of receptors (Evans 1988) and has been characterized by molecular biological analyses in great detail. It exhibits a characteristic domain structure which comprises an N-terminal modulator region, a central DNA binding region, and a C-terminal hormone binding region (Hollenberg et al. 1987; Hollenberg and Evans 1988).

[1]Friedrich Miescher-Institut, P.O. Box 2543, CH-4002 Basel, Switzerland

40. Colloquium Mosbach 1989
Molecular Mechanisms of Hormone Action

2.1.1 Glucocorticoid Receptor Levels Are Down-Regulated upon Hormone Activation

We have followed the fate of the GR intracellularly, upon hormone addition to the growth medium of cultured NIH 3T3 fibroblasts. Two methods were used to measure the GR concentration as a function of time after hormonal induction. The binding to whole cells of [^{3}H]-triamcinolone acetonide was assayed (Pfahl et al. 1978; McIntyre and Samuels 1985) and indicated that the binding potential for this radioactive glucocorticoid is maximal at time zero of hormone addition and decreases steadily to about 35% of the maximal value 17 h later (Hoeck et al. 1989). To distinguish between a functional modification of the GR and a decrease in its concentration, a specific antiserum was used to determine the GR concentration in protein blot experiments. Nuclear and cytoplasmic receptors were analyzed and it was found that the cytoplasmic form translocated nearly quantitatively to the nucleus within 30 min after hormone addition and that the nuclear receptor decayed to about 20% of its initial, maximal level after 24 h of hormone induction. These findings indicate that ligand addition results in a down-regulation of nuclear receptor levels due to degradation of the translocated molecules. The low glucocorticoid receptor level observed after 24 h of hormone induction remained stable during longer periods of hormone exposure. These residual receptor molecules might be important in the negative regulation of the GR gene as has been speculated by Okret et al. (1986). Removal of the hormone from the growth medium resulted in a recovery to higher receptor concentrations. Half-maximal levels were observed 12 h after hormone withdrawal and maximal levels were restored after 24 h of growth in hormone-free medium (Höck et al. 1989a).

2.1.2 GR Down-Regulation is Due to a Decrease of GR mRNA and GR Protein Half-Life

The decrease of the cellular GR concentration was caused by different molecular mechanisms. A mouse GR gene-specific probe was used to quantitate GR mRNA levels at various times after hormone induction of NIH 3T3 cells. An RNA blot analysis revealed a rapid decrease in the concentration of the 4.8 and 6.5 kb GR-specific mRNA. Only 30% of the maximal mRNA level found before hormone addition was present in cells treated for 3 h with hormone. Longer hormone treatment, up to 24 h, did not result in a further decrease in GR mRNA concentration. This experiment indicates that a decrease in mRNA and subsequently a decreased rate of GR protein synthesis contributes to the down-regulation process (Rosewicz et al. 1988; Dong et al. 1988; Kalinyak et al. 1987; Höck et al. 1989b).

A second level of regulation was discovered when the half-life of GR was determined in the absence and presence of hormone. Cells were pulsed with [^{35}S]-methionine for 16 h and GR was immunoprecipitated after various times following replacement of the radioactive amino acid with nonradioactive methionine. The radioactivity associated specifically with the 100 kDa GR was determined and indicated a GR half-life of 8 h in noninduced cells. The half-life was reduced to 3 h in hormone-treated cells (McIntyre and Samuels 1985; Höck et al. 1989a). The reduced receptor half-life can be due either to a destabilization of GR after hormone activation or to an increased efficiency of the degradation process. These experiments showed that a reduced rate of synthesis and an enhanced rate of degradation simultaneously lower the GR concentration in hormone-stimulated cells.

2.1.3 GR is Rapidly Phosphorylated upon Hormone Activation

Steroid receptor molecules are phosphoproteins (Housley and Pratt 1983) but the role of phosphorylation in receptor function or receptor down-modulation has remained elusive. If receptor phosphorylation is functionally significant, a change in the extent or pattern of phosphorylation upon hormone induction and nuclear translocation would be expected. We have used immunoprecipitation of receptor labeled with [^{32}P]-orthophosphate in vivo to address this question. Autoradiography of nuclear receptor immunoprecipitated 0, 20, 40 and 60 min after hormone addition revealed that the specific activity increased about fourfold within 40 min (Höck et al. 1989a). This increase in phosphorylation coincides with the transcriptional activation of glucocorticoid responsive genes and precedes the down-regulation described above. A phosphoamino acid analysis of the modified amino acids in the immunoprecipitated GR revealed mostly phosphoserine residues.

We observed earlier that the glucocorticoid antagonist RU 38486 inhibited hormone-dependent transcription rapidly upon addition to the medium of mouse fibroblasts (Groner et al. 1983). We analyzed the effect of RU 38486 on the extent of receptor phosphorylation. RU 38486 caused nuclear translocation of the GR with a reduced efficiency (W. Höck, unpublished observations). The translocated GR did not show an increase in phosphorylation when compared to the nonactivated cytoplasmic form. This experiment indicates that receptor phosphorylation might be required for the functional activation in the nucleus.

2.1.4 Ha-ras and v-mos Oncogene Expression Enhances GR Down-Regulation

We previously observed that the glucocorticoid hormone-induced transcription, for instance, of the MMTV LTR in NIH 3T3 cells reaches a maximum 30 min after hormone addition. We also observed that prolonged exposure of cells to hormones causes a reduction of the maximal transcriptional rate. Fifty to 60% of the maximal rate was measured after 24h of hormonal induction. We compared the transcriptional rates at various times after hormonal stimulation in NIH 3T3 cells and NIH 3T3 cells expressing the Ha-ras or v-mos oncogenes. We found that the decrease in the transcription rate occurring at prolonged hormone exposure periods was accentuated more in the oncogene-expressing cells (Jaggi et al. 1986, 1989). The decrease in nuclear GR concentration described above correlates well with the change in transcription activity in normal NIH 3T3 cells. In order to test if the effect of oncogenes on hormonally-induced gene transcription is mediated by GR concentrations, we compared the nuclear GR concentrations in normal and transformed cells at different times after hormone addition. We found that the down-regulation of GR is enhanced in oncogene-expressing cells (Höck et al. 1989b; Jaggi et al. 1989), i.e., the observations made earlier on the repression of glucocorticoid-dependent gene transcription by oncogenes is effected via the accelerated degradation of the GR in transformed cells.

Figure 1 summarizes some of the experimental results described above. Phosphorylation of the GR is a very early event upon nuclear translocation. It is concurrent with or precedes transcriptional activation, for instance, of the MMTV LTR and transcriptional repression, for instance, of the GR gene. The nuclear localization

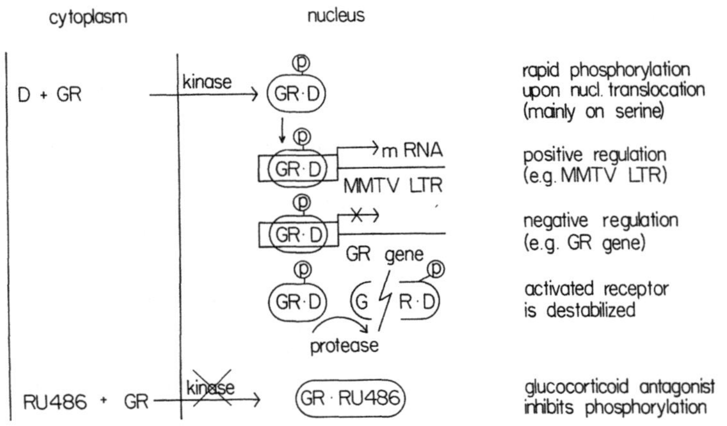

Fig. 1. Fate of the glucocorticoid receptor upon hormone activation

results in a decreased stability of the GR. RU 38486 prevents GR phosphorylation and transcriptional transactivation, but might not interfere with the down-regulation process.

2.2 Glucocorticoid Hormone Action Provides Competence to Mammary Epithelial Cells to Respond to Prolactin

The remarkable progress in the molecular biology of steroid hormone receptors narrowed the endocrinological outlook somewhat at the expense of hormonal physiology. The cooperation of different classes of hormones, the dependence of the potential of individual hormones on cell type and cellular environment are complex questions not yet accessible to the high resolution of molecular biological methods. The regulation of differentiation and gene expression in mammary epithelial cells is an important example of hormonal synergism in the control of physiological processes (Topper and Freeman 1980). The accessibility was limited in the past to in vivo studies and primary cell explants (Lee et al. 1985). Only recently, advances in the establishment of cell lines and tissue culture techniques have made it possible to approach these questions with recombinant DNA techniques and gene transfer methods.

2.2.1 Glucocorticoid Hormone and Prolactin Act Synergistically in the Induction of the β-Casein Gene Promoter in Vitro

An important advance in the studies of hormonal regulation of milk protein gene expression was the establishment of a clonal cell line from the mammary tissue of mid-pregnant mice (Danielson et al. 1984; Ball et al. 1988). This cell line proved to be independent of added extracellular matrix and heterologous cell interactions in its potential to respond to lactogenic hormones and synthesize milk proteins. Transcriptional regulation of the β-casein gene promoter by the action of insulin, dexamethasone, and prolactin was established in nuclear run-on experiments (Ball et al. 1988). This cell line (HC11) is also transfectable. The responsiveness to the lactogenic hormones on the transcriptional level and the possibility to introduce in vitro recombined genes allowed the application of gene transfer techniques to the question of hormonal synergism. β-Casein promoter-CAT gene constructs were introduced stably into HC11 cells. CAT activity was measured in cell extracts after different hormone treatment schedules. Prolactin or dexamethasone administration alone had only very minor effects on the enzyme activity. Only a combination of the two hormones was able to induce high levels of CAT activity. Furthermore, hormonal induction was restricted to confluent cell cultures (Doppler et al. 1989). These results revealed two important prerequisites for activation of the β-casein promoter: (1) HC11 cells must reach quiescence, and (2) signals from the glucocorticoid receptor and the prolactin receptor must synergize.

2.2.2 Glucocorticoid Hormone and Prolactin Act Sequentially in the Induction of the β-Casein Gene Promoter

A simple model of hormonal synergism might postulate a sequential or simultaneous mode of action. It is conceivable that both types of signals, i.e., those generated by the activated GR and those generated by the activated prolactin receptor yield a functional activation of transcription factors which regulate the β-casein gene promoter. Alternatively, it is possible that other cellular genes are involved and that one or both hormones regulate the β-casein promoter in an indirect fashion. Hormone pretreatment experiments were carried out, i.e., the hormones were added sequentially to confluent β-casein-CAT transfected HC11 cells and the time course of CAT activity accumulation was determined. Pretreatment of cells with prolactin did not accelerate the rate of CAT activity accumulation upon addition of dexamethasone when these cells were compared to nonpretreated control cells induced with prolactin and dexamethasone simultaneously. Pretreatment of cells with dexamethasone, on the contrary, had a very marked effect. The longer the pretreatment was carried out (up to 15 days), the more efficiently the cells responded to prolactin. The acquisition of competence to respond to prolactin was therefore totally dependent on the action of dexamethasone. This process was slow, i.e., a maximal effect of dexamethasone had still not been reached after 14 days.

Withdrawal of dexamethasone after 6 days of pretreatment revealed that the dexamethasone conferred competence is a rapidly reversible process. After 1 day of withdrawal the prolactin-induced CAT activity approaches that of nonpretreated cells.

2.2.3 The Sequences Required for the Synergistic Induction of the β-Casein Gene Promoter are Contained Within 330 Nucleotides of the 5′ Flanking Region

The hormonal inducibility of β-casein promoter-CAT constructs was initially tested with in vitro recombined genes which contained about 2.3 kb of β-casein gene DNA 5′ of the RNA initiation site (Doppler et al. 1989). A sequence delimitation analysis from the 5′ end was carried out and each deletion mutant was tested for its ability to respond to the lactogenic hormones in stably transfected HC11 cells. We found that 330 nucleotides were sufficient to confer hormonal inducibility. A comparison of the extent of hormonal induction revealed that several sequence elements contribute to the overall strength of the β-casein promoter. Although inducibility was retained in deletions to about −180, the induction factor decreased (Doppler et al. 1989). It is conceivable that qualitative, i.e., elements which are hormonally sensitive, and quantitative, i.e., elements which modulate the promoter strength, interact in the regulation process.

2.2.4 A Sequence of Negative and Positive Regulation Could Explain the Hormonal Synergism

The experiments described above clearly indicate that dexamethasone and prolactin do not act simultaneously on their induction on the β-casein promoter. A perplexing feature of the conferral of cellular competence to react to prolactin by dexamethasone is the very slow time course. It takes at least 14 days in nondividing, confluent HC11 cells before the maximal effect is reached, i.e., before the cells are maximally sensitive to the action of prolactin. A hypothetical model is shown in Fig. 2 which accommodates these observations and which might explain the molecular connections. We speculate that the β-casein gene is regulated by positive as well as negative signals. In the differentiated cells it is conceivable that a repressor protein (RP) is synthesized constitutively. This protein interacts with the β-casein promoter sequence and blocks transcription. The negative regulation is exerted as long as RP is present. GR has been shown to regulate genes positively and negatively (Beato 1989) and we speculate that the RP gene is subject to negative control by the GR. Addition of glucocorticoid hormone to HC11 cells might result in a repression of the RP gene. This would not result in an instantaneous inducibility of the β-casein gene promoter if the RP has a long half-life, i.e., a high stability. Only the decay of the RP results in the competence of the HC11 cells to respond to prolactin. Our model predicts a prolactin response element in the β-casein gene promoter. This prl RE is the binding site for a transcription factor which is activated, perhaps by secondary modification, through the

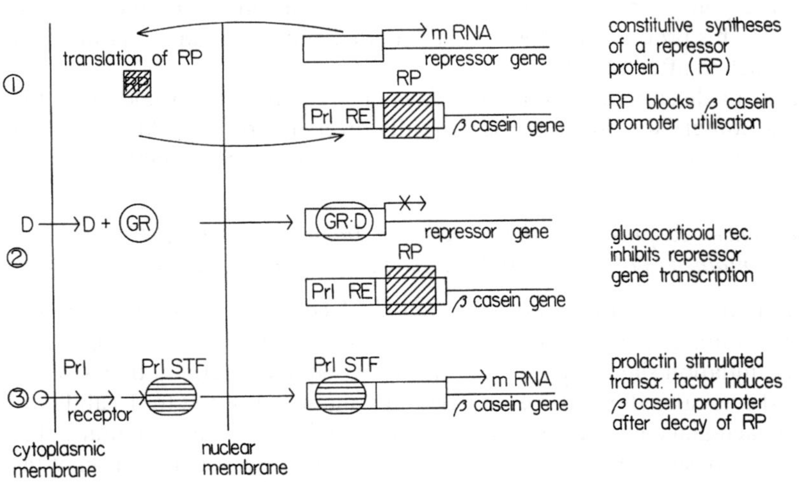

Fig. 2. Model for the sequential action of dexamethasone and prolactin on the β-casein gene induction in HC11 cells

action of a ligand-activated prolactin receptor. The prolactin stimulated transcription factor (prl STF) can only exert its action in the absence of RP, i.e., RP action is dominant over prl STF function. A detailed analysis of the *cis*-acting DNA sequence elements contained in the β-casein gene promoter and an identification of the transacting transcription factors interacting with these regulatory DNA sequences will be required to test this model.

References

Ball, R., Friis, R., Schönenberger, C.A. & Groner, B. (1988) Prolactin regulation of β-casein gene expression and of a 120 kDa protein in a cloned mouse mammary epithelial cell line. EMBO J. 7:2089–2095

Beato, M. (1989) Gene regulation by steroid hormones. Cell 56:335–344

Danielson, K.G., Oborn, C.J., Durban, E.M., Butel, J.S. & Medina, D. (1984) Epithelial mouse mammary cell line exhibiting normal morphogenesis in vivo and functional differentiation in vitro. Proc. Natl. Acad. Sci. USA 81:3756–3760

Danielson, M., Northrop, J.P. & Ringold, G.M. (1986) The mouse glucocorticoid receptor: mapping of functional domains by cloning, sequencing and expression of wild type and mutant receptor proteins. EMBO J. 5:2513–2522

Dong, Y., Poellinger, L., Gustafsson, J.A. & Okret, S. (1988) Regulation of glucocorticoid receptor expression: evidence for transcriptional and posttranscriptional mechanisms. Mol. Endocrinol. 2:1256–1264

Doppler, W., Groner, B. & Ball, R. (1989) Prolactin and glucocorticoid hormones synergistically induce expression of transfected rat β-casein gene promoter constructs in a mammary epithelial cell line. Proc. Natl. Acad. Sci. USA 86:104–108

Evans, R.M. (1988) The steroid and thyroid hormone receptor superfamily. Science 240:889–895

Green, S. & Chambon, P. (1988) Nuclear receptors enhance our understanding of transcriptional regulation. Trends Genet. 4:309–314

Groner, B., Hynes, N.E., Rahmsdorf, U. & Ponta, H. (1983) Transcription initiation of transfected mouse mammary tumor virus LTR DNA is regulated by glucocorticoid hormones. Nucl. Acids Res. 11:4713–4725

Höck, W., Rusconi, S. & Groner, B. (1989a) Downregulation and phosphorylation of glucocorticoid receptors in vivo: investigations with a monospecific antiserum against a bacterially expressed receptor fragment. Biol. Chem. 264:No 24

Höck, W., Pfahl, M., Jaggi, R. & Groner, B. (1989b) Ligand induced downregulation of glucocorticoid receptors is enhanced by the expression of oncogenes. In: Lippman, M. & Dickson, R. (eds) Growth regulation of cancer. II. Liss, New York

Hollenberg, S.M. & Evans, R.M. (1988) Multiple and cooperative trans-activation domains of the human glucocorticoid receptor. Cell 55:899–906

Hollenberg, S.M., Giguére, V., Segui, P. & Evans, R.M. (1987) Co-localisation of DNA binding and transcriptional activation functions in the human glucocorticoid receptor. Cell 49:39–46

Housley, P.R. & Pratt, W.B. (1983) Direct demonstration of glucocorticoid receptor phosphorylation by intact L-cells. J. Biol. Chem. 258:4630–4635

Jaggi, R., Salmons, B., Müllener, D. & Groner, B. (1986) The v-mos and H-ras oncogene expression represses glucocorticoid hormone dependent transcription from the mouse mammary tumor virus LTR. EMBO J. 5:2609–2616

Jaggi, R., Höck, W., Ziemiecki, A., Klemenz, R., Friis, R. & Groner, B. (1989) Oncogene mediated repression of glucocorticoid response elements and glucocorticoid receptor levels. Cancer Res. (in press)

Kalinyak, J.E., Dorin, R.I., Hoffman, A.R. & Perlman, A.J. (1987) Tissue specific regulation of glucocorticoid receptor mRNA by dexamethasone. J. Biol. Chem. 262:10441–10444

Lee, E.Y., Lee, H.P., Kaetzel, C.S., Parry, G. & Bissell, M.J. (1985) Interaction of mouse mammary epithelial cells with collagen substrata: regulation of casein gene expression and secretion. Proc. Natl. Acad. Sci. USA 82:1419–1432

McIntyre, W.R. & Samuels, H.H. (1985) Triamcinolone acetonide regulates glucocorticoid receptor levels by decreasing the half-life of the activated nuclear receptor form. J. Biol. Chem. 260:418–427

Okret, S., Poellinger, L., Dong, Y. & Gustafsson, J.A. (1986) Downregulation of glucocorticoid receptor mRNA by glucocorticoid hormones and recognition by the receptor of a specific binding sequence within a receptor cDNA clone. Proc. Natl. Acad. Sci. USA 83:5899–5903

Pfahl, M., Sandros, T. & Bourgeois, S. (1978) Interaction of glucocorticoid receptors from lymphoid cell lines with their nuclear acceptor sites. Mol. Cell. Endocrinol. 10:175–191

Rosewicz, S., McDonald, A.R., Maddux, B.A., Goldfine, I.D., Miesfeld, R.L. & Logsdon, C.D. (1988) Mechanism of glucocorticoid receptor downregulation by glucocorticoids. J. Biol. Chem. 263:2581–2584

Topper, Y. & Freeman, C.S. (1980) Multiple hormone interactions in the developmental biology of the mammary gland. Physiol. Rev. 60:1049–1106

Studies on the Regulation of Glycogen and Lipid Metabolism by Insulin and Growth Factors: The Involvement of Receptor Tyrosine Kinase Activation and Casein Kinase II

T.A.J. HAYSTEAD[1] and E.G. KREBS[1]

1 Introduction

One of the fundamental questions today in studies on the regulation of glycogen and lipid metabolism by insulin and the growth factors whose receptors are protein tyrosine kinases, is how do these ligands on binding to their respective receptors communicate responses to the cytoplasm. Initial events in signaling begin with increased autophosphorylation on tyrosine and serine residues of the cytoplasmic domains of the receptors themselves. Indeed, it is now generally accepted that binding of these hormones results in the activation of intrinsic tyrosine kinase activity, and that this activity is prerequisite and essential for all subsequent metabolic events (for reviews, see Denton 1986; Rosen 1987; Pelech et al. 1987; Czech et al. 1988). These immediate events bring about both increases and decreases in the phosphorylation state (on serine and threonine residues) of many of the important regulatory proteins and enzymes of protein synthesis, glycogen and lipid metabolism, e.g., ribosomal protein S6, glycogen synthase, pyruvate-dehydrogenase, ATP-citrate lyase, acetyl-CoA carboxylase, hormone-sensitive lipase. Importantly, in some instances, good evidence has been obtained to show that these effects on phosphorylation correlate well with changes in the flux of substrates through each respective pathway and also with changes in the intrinsic activity of the individual regulatory steps concerned (for reviews, see Strålfors et al. 1984; Cohen 1986; Roach 1986; Denton 1986).

Recent studies by this laboratory and others, using isolated cells in culture, have demonstrated that growth factor stimulation rapidly activates several protein serine/threonine kinases (Cobb and Rosen 1982; Novak-Hofer and Thomas 1985; Erikson and Maller 1986; Sommercorn et al. 1987; Giugni et al. 1988; Klarlund and Czech 1988). Examples of the activated kinases that have been identified are casein kinase II (CKII) and ribosomal S6 kinase. For the S6 kinase, activation appears to involve serine/threonine phosphorylation, since serine/threonine and tyrosine protein phosphatase inhibitors are required to observe and maintain the activated state, both in crude cell extracts prepared from stimulated cells and during subsequent purification steps (Cobb and Rosen 1982; Novak-Hoffer Thomas 1985; Jenö et al. 1989). Ray and Sturgill (1987, 1988a,b) recently tentatively identified MAP-2 kinase (myelin associated basic protein kinase) as the ribosomal S6 kinase kinase responsible for the activation of S6 kinase. Indeed, they demonstrated that this "kinase kinase" was activated by growth factor treatment and that this activation was itself was associated with increases in tyrosine and threonine phosphorylation on the enzyme. Activation of

The Howard Hughes Medical Institute, University of Washington, Seattle, WA 98195, USA

40. Colloquium Mosbach 1989
Molecular Mechanisms of Hormone Action

CKII by these hormones would also appear to involve some form of covalent modification since the hormone-stimulated activity is stable to high dilution and partial fractionation (Sommercorn et al. 1987). Moreover, it has been determined that treatment of activated CKII by alkaline phosphatase returns the enzyme to a basal activity state (N. Osheroff, personal communication).

Simplistically, it would seem possible that activation of tyrosine kinases by hormones and growth factors might involve a mechanism of signal transduction in which the phosphorylation on tyrosine residues of serine/threonine protein kinases would bring about their activation. Such a mechanism is certainly plausible in the cases cited above, although direct evidence has not yet been obtained to demonstrate that these kinases are in fact activated directly by tyrosine phosphorylation. In the case of the MAP-2 kinase the increased tyrosine phosphorylation implies that this is the activating mechanism, although this remains to be more fully characterized, since there was also increased threonine phosphorylation in response to the growth factor (Ray and Sturgill 1988b). Alternative hypotheses have suggested, at least in the case of insulin, that an intracellular mediator is released from the membrane which then results in the modulation of protein kinase and/or protein phosphatase activity. Evidence for such a mediator has been obtained by Larner and co-workers and later by Saltiel and others (Larner et al. 1979; Jarett et al. 1985; Saltiel et al. 1986; Kelly et al. 1987). The mediator has been identified as a phosphoinositol-glycan released from the membrane by the activation of a specific phospholipase C, which subsequently cleaves the phosphoinositol-glycan from the membrane. In this way it has been suggested that tyrosine kinase activation is requisite for the allosteric interaction of the receptor with the phosphoinositol-glycan effector system, which then modulates all subsequent steps. Actual evidence for this hypothesis has not yet been obtained. It may be that direct modulation of protein kinases by tyrosine phosphorylation is a separate event occurring in conjunction with phosphoinositol glycan generation.

It was mentioned above that growth factor treatment of isolated cells also brings about dephosphorylation of some key regulatory enzymes, examples of which include glycogen synthase, pyruvate dehydrogenase (PDH), acetyl-CoA carboxylase, and hormone-sensitive lipase (Parker et al. 1983; Strålfors et al. 1984; Reed and Yeamen 1986; Munday and Hardie 1986; Witters et al. 1988). These changes in activity correlate well with decreased phosphorylation of the respective enzymes, but the mechanism by which the phosphatases involved might be activated remains to be elucidated. With perhaps the exception of the mitochondrial PDH protein phosphatase, direct evidence that the protein phosphatases are activated has been lacking. However, recently this laboratory described for the first time the direct activation of a type 1 protein phosphatase activity (PP-1) in 3T3-D1 cells in response to physiological doses of insulin, EGF, and PDGF which was measurable in extracts from the treated cells.

In the 3T3 cell system, there is both activation of protein serine/threonine kinases and a protein serine/threonine phosphatase in response to insulin and growth factor. The functional roles of these activations is still unclear, since with the possible exception of ribosomal protein S6 kinase (Palen and Traugh 1987), the physiologically significant targets of these activated enzymes remain to be defined. Activation of CKII is of particular interest, since in vitro and in vivo studies have shown this protein kinase phosphorylates several important enzymes involved in regulating the metabolism of glycogen and lipid, examples of which include glycogen synthase (Parker et al. 1983) and acetyl-CoA carboxylase (Haystead et al. 1988). The kinase has also been implicated in the activation of a cytosolic form of PP-1, PP-1_I or so-called Mg/ATP-dependent protein phosphatase (Goris et al. 1979; Yang et al. 1980, 1982; Hemmings et al. 1982; DePaoli-Roach 1984; Ballou and Fischer 1986). It is the purpose of this chapter to define more fully the possible role(s) of CKII activation by growth factor stimulation. The possible involvement of CKII with protein phosphatase 1 and how this may be related to the regulation of metabolism in 3T3 cells will be discussed.

2 Results and Discussion

2.1 Hormonal Modulation of Glycogen Synthase Activity

Earlier, it was reported by this laboratory that addition of physiological doses of EGF (10 nM), insulin (10 nM), or PDGF (5 ng ml^{-1}) to quiescent 3T3 cells brought about a maximal 2.5-fold activation of glycogen synthase activity within 30 min. This activation was followed by a refractory period in which synthase activity returned to basal levels (Fig. 1). Direct quantization of known allosteric effectors (glucose 6-phosphate, adenine nucleotides, and Pi) in the enzyme assay mixtures indicated that the observed activation was not simply a consequence of changes in metabolite concentrations (Chan and Krebs 1985; Chan et al. 1987). These findings suggest that like insulin, EGF and PDGF may be important factors to consider in the regulation of glycogen synthase activity. They also add to the growing body of evidence that insulin and growth factors act through common mechanisms of signal transduction.

Although glycogen synthase is subject to modulation by allosteric metabolites, phosphorylation/dephosphorylation in response to hormonal stimulation is now

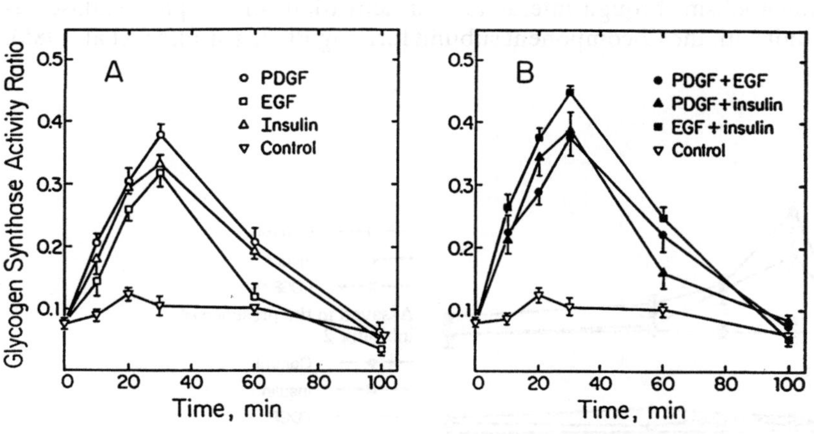

Fig. 1. Effect of insulin (100 ng ml^{-1}), EGF (100 ng ml^{-1}), and PDGF (5 ng ml^{-1} on glycogen synthase activity in 3T3 cells. Confluent quiescent cultures of 3T3 cells were incubated with DME medium either with BSA (0.1 mg ml^{-1}), insulin, EGF, or PDGF either alone or in combination. **A** Factors added alone; **B** factors added in combination. Values plotted are means ± SEM (n = 3 plates) from a single experiment. The same experiment has been repeated three times with very similar results. These data were previously presented by Chan et al. (1987)

generally believed to have the most significant effect on the regulation of enzyme activity in vivo. Dephosphorylation, in response to insulin, brings about activation and increased phosphorylation, in response to adrenaline or glucagon, brings about inactivation (Lawrence and Hinken 1983; Parker et al. 1983; Roach 1986; Cohen 1986). Although the purified enzyme contains in vitro phosphorylation sites for at least ten protein kinases, it is thought that only a few of these have importance in vivo in the activation/inactivation of the enzyme. Of major significance in the regulation of activity are three closely grouped phosphorylation sites in the C terminal domain (sites 3, serines 30, 34, and 38 of cyanogen bromide peptide CB-C, see Cohen 1986). These are exclusively phosphorylated by GSKIII (glycogen synthase kinase III), resulting in inhibition of activity. Phosphorylation of these sites and also serine 42 by GSKIII requires prior phosphorylation of serine 46 by CKII, an event which is essential to bring about inhibition of activity (DePaoli-Roach 1984; Fiol et al. 1987). Significantly, CKII phosphorylation alone does not appear to effect enzyme activity directly.

The mechanism by which glycogen synthase is activated and inactivated is some what unclear, since phosphorylation of sites 3 brings about inhibition, and insulin has been shown to promote the dephosphorylation of these sites (Parker et al. 1983; Cohen 1986). Therefore, the synergistic relationship between GSKIII and CK II phosphorylation represents a paradoxical situation, particularly in the 3T3 cell, in the light of findings presented above, i.e., activation of CKII by insulin or growth factor would in theory lead to inhibition of glycogen synthase activity through its synergistic relationship with GSKIII. The situation is further complicated in the 3T3 cell by the observation that PP-1 is also activated by these hormones (Fig. 2), an event which could possibly lead to synthase dephosphorylation and activation. Therefore, the question is raised, are these related or separate events in terms of the regulation of glycogen synthase activity in 3T3 cells?

In muscle, and possibly liver, Cohen and co-workers (1986) have postulated that insulin would promote association of PP-1 with glycogen and the enzymes of glycogen metabolism through interaction (or activation) of the phosphatase catalytic subunit with the G-component subunit forming PP-1_G (Strålfors et al. 1985).

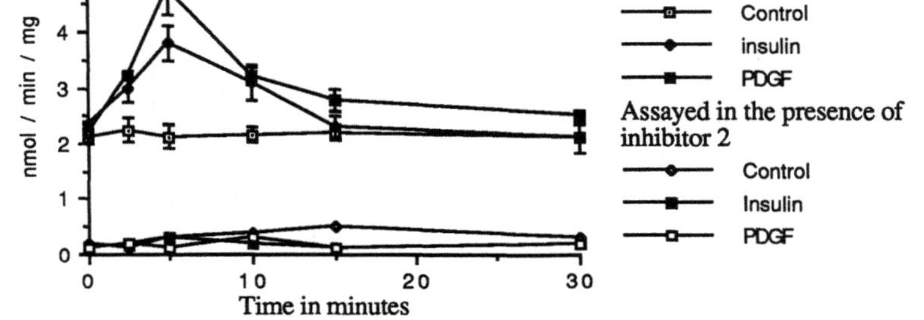

Fig. 2. Time course of the effect of insulin and PDGF on protein phosphatase 1 activity. Confluent quiescent cultures of 3T3 cells were incubated at 37°C in DME medium, 2% CMS-1 (Ross et al. 1978) with either no hormone, PDGF (1 ng ml^{-1}) or insulin (100 ng ml^{-1}). Values are ± SEM (n = 3) from three separate experiments (see also Chan et al. 1988)

Evidence for such a reassociation, or activation of bound PP-1, has not yet been obtained. However, these workers have obtained good evidence in muscle that demonstrates the translocation of PP-1 catalytic subunit from glycogen associated PP-1_G to the cytosol, with concomitant increased phosphorylation of the G subunit in response to adrenergic stimulation (Hiraga and Cohen 1986; MacKintosh et al. 1988). Earlier, in vitro studies demonstrated that the G subunit could be phosphorylated by cyclic-A kinase (cyclic-AMP dependent protein kinase) and that this also caused dissociation of PP-1_G catalytic subunit from the glycogen particle (Caudwell et al. 1986).

2.2 *Mg/ATP-Dependent Protein Phosphatase (PP-1_I)*

In 3T3 cells it would seem that the form of PP-1 responsible for the dephosphorylation and activation of glycogen synthase in response to growth factors would be the form that has actually been shown to undergo activation under these circumstances (Chan et al. 1988). This phosphatase was found to be activated within the same time frame as the activation of glycogen synthase (Fig. 2; see also Chan et al. 1988). The possible identity of this protein phosphatase has been recently characterized by this laboratory and our current hypothesis is that it is an activated form of PP-1_I, i.e., the so-called Mg/ATP-dependent phosphatase as originally described by Goris et al. (1979). Evidence for this hypothesis comes from recent findings that this form of PP-1 accounts for the major protein phosphatase activity in 3T3 cells (Fig. 3). Furthermore, no evidence has been obtained showing the presence of glycogen-bound or other bound forms of PP-1 as defined by Cohen and co-workers (Stralfors et al. 1985; Hiraga and Cohen 1986; Chisholm and Cohen 1988). As shown in Figs. 2 and 3, the activated PP-1 was present in high-speed supernatent extracts (100 000 × g) prepared from

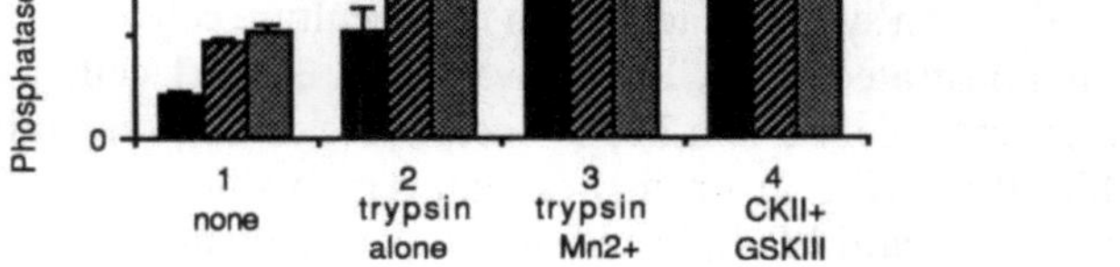

Fig. 3. The presence of protein phosphatase 1_I (PP-1_I) in 3T3-D1 cells. High-speed supernatent cell extracts (100 000 × g) were prepared in the absence of protease inhibitors from cultures incubated with either no hormone, insulin (100 ng ml^{-1}), or PDGF (1 ng ml^{-1}) for 5 min at 37°C, then freeze clamped by floating culture dishes on ethanol/dry ice. For measurements of phosphatase activity, samples containing 8 μg of extract protein were preincubated in a solution of 20 mM imidazole (pH 7.4), 5 mM 2-mercaptoethanol, plus the indicated components for 5 min at 30°C prior to the addition of ^{32}P phosphorylase *a* solution, containing 20 μg ml^{-1} lima bean trypsin inhibitor and 10 mM EDTA. Tryptic (10 μg ml^{-1}) digestion was carried out in the presence or absence of 1 mM Mn^{2+}. Purified preparations of CKII (final dilution 10 U ml^{-1}) and GSKIII (final dilution 12.5 U ml^{-1}) were added in the presence of 1 mM ATP, 25 mM MgCl. Values are means ± SEM from three separate experiments

stimulated cells. Little, or no activity, was detectable in the particulate fractions prepared from either nonstimulated, or hormone-stimulated cell extracts. In the case of muscle, or liver, significant proportions of the total cellular PP-1 can be accounted for as being bound forms, either associated with myosin or the glycogen particle (Cohen 1986; Chisholm and Cohen 1988).

Isolated in pure form PP-1_I is completely inactive, has a molecular mass of 70 000 kDa and consists of a complex of two separable subunits, the heat-stable inhibitor 2 protein (Mr 23 000) and the free catalytic subunit (Mr 33 000) (Ballou et al. 1983; Holmes et al. 1986a). The inactive complex can be activated in two characteristic and diagnostic ways, either by brief trypsin and Mn^{2+} exposure (Fig. 3; Brautigan et al. 1980, 1982) or by incubation with Mg/ATP in the presence of GSKIII (Fig. 3; Goris et al. 1979; Hemmings et al. 1982; Jurgenson et al. 1984; Roach 1984). Using the purified enzyme, after phosphorylation by GSKIII, Moruzzi et al. (1984) were able to isolate various activated forms of PP-1_I and PP-1_I subunits on FPLC (Fast Protein Liquid Chromatography) anion-exchange chromatography. Their findings enabled them to propose a complex model for the activation and inactivation of PP-1_I involving phosphorylation of the inhibitor 2 molecule by GSKIII, while complexed with the catalytic subunit (see Ballou and Fisher 1986). It was demonstrated that the inhibitor 2 molecule was essential for both the activation and inhibition of the complex and represents a unique modulator in metabolism. Interest was further stimulated by the observation of DePaoli-Roach (1984) that phosphorylation of the inhibitor 2 molecule by CKII on serines 120, 121, and 86 (Holmes et al. 1986b, 1987) was prerequisite to phosphorylation by GSKIII on threonine 72 (Holmes et al. 1986b). The regulation of PP-1_I activity represents yet another example of a synergistic relationship between CKII and GSKIII in the regulation of enzyme activity. Earlier, it was reported by this laboratory that insulin promoted the activation of CKII in 3T3-L1 cells (Sommercorn et al. 1987). The activation of CKII would therefore afford a means for the activation of PP-1_I, although this hypothesis should not exclude the possibility of GSKIII activation (see Yang et al. 1988).

2.3 CKII and the Regulation of Glycogen Metabolism in 3T3 Cells

Taking existing observations, it may be possible to explain the effects of prolonged growth factor stimulation on glycogen synthase activity in tissue culture cells (Fig. 1). Sommercorn et al. (1987) demonstrated that CKII activation in 3T3-L1 cells was maximal as early as 10 min. Activation of PP-1 in 3T3-D1 cells is maximal after 5 min, returning to basal levels within 20 min (Fig. 2). From these data it is possible to propose that CKII activity is first targeted toward PP-1_I, resulting in its subsequent activation by GSKIII. Activated PP-1_I could then bring about dephosphorylation of glycogen synthase (and possibly other substrates). As CKII continues to be activated, glycogen synthase becomes targeted and susceptible to GSKIII phosphorylation, resulting in an antagonistic inhibitory effect. In order to provide evidence for such a cause and effect hypothesis in 3T3 cells, the relative rates of activation of individual protein kinases involved and the phosphorylation of the respective sites on PP-1_I and glycogen synthase must be determined. It should be stressed also that, although the observations of CKII and PP-1 activation were both made in 3T3 cells in response to growth factor stimulation, these phenomena were in different 3T3 cell lines, i.e., transformed 3T3-L1 for CKII, and nontransformed 3T3-D1 for PP-1. There is a need, therefore, to correlate both activities in at least one of these lines. Preliminary findings by this laboratory indicate that CKII activation does indeed closely follow PP-1 activation in 3T3-D1 cells following PDGF stimulation (Haystead and Litchfield unpublished observations).

The physiological significance of these findings may only apply to the regulation of glycogen metabolism in 3T3 cells, since other mechanisms may be in operation in other cell types, such as muscle, for example, where translocation of phosphatase activity to and from various subcellular locations may be important. However, clearly in the study of signal transduction, CKII activation in the 3T3 cell represents a useful model for the elucidation of the mechanisms by which not only activated receptor tyrosine kinases may communicate with protein serine/threonine kinases, but also how these activated protein kinases can subsequently modulate regulatory steps in metabolism. Findings such as these have helped to reenforce the physiological significance of CKII activity.

2.4 Other Functions of CKII

One of the common features of phosphorylation of substrates by CKII is that it does not ordinarily appear to bring about direct changes in enzymatic activity (with the possible exception of topoisomerase II, Ackerman et al. 1985, 1988). This has cast doubt on the importance of this protein kinase. Phosphoserine has been detected in the CKII sites of several important regulatory molecules, including the inhibitor 2 molecule, RII (the regulatory subunit of cyclic-A kinase), glycogen synthase, and acetyl-CoA carboxylase. From studies carried out in muscle by Cohen and co-workers, the phosphorylation of these sites does not appear to change in response to hormonal stimulation (Parker et al. 1983; Cohen 1986; Holmes et al. 1986b). Indeed, to date acetyl-CoA carboxylase represents the only case reported where phosphorylation of CKII sites are increased in response to hormonal stimulation (Witters et al. 1983; Holland and Hardie 1985; Haystead et al. 1988). There is a need, therefore, to look more closely at these other substrates in order to ascertain further the role of CKII (if any) in the regulation of these enzymes and their respective pathways. It is difficult to imagine that these sites are merely redundant or "silent". Clearly the synergistic relationship of CKII and GSKIII in the phosphorylation of some of these substrates implicates a function for CKII.

With acetyl-CoA carboxylase, activation of the enzyme is response to insulin does not seem to be directly related to increased phosphorylation in its CKII phosphorylation sites, since the activated state (observed in cell extracts) is not maintained through purification to homogeneity, unlike the effects of the hormone on phosphorylation. In these experiments, isolated cells (hepatocytes and adipocytes) were treated with insulin for a maximum of 15 min (Witters et al. 1983; Holland and Hardie 1985; Haystead et al. 1986, 1988). It is possible, however, that CKII phosphorylation may target the enzyme for some later modification. One potential modification would be dephosphorylation, an event which is known to bring about activation (Haystead and Hardie 1986; Munday and Hardie 1986; Witters et al. 1988). Evidence for such a role for CKII in the regulation of acetyl-CoA carboxylase activity has come from observations made by this laboratory. Sommercorn and Krebs (1987) were able to demonstrate that in vitro phosphorylation of acetyl-CoA carboxylase by CKII rendered the enzyme increased in its susceptibility to dephosphorylation at its inhibitory phosphorylation site (the in vitro phosphorylation site of cyclic-A kinase, which is now known to be phosphorylated by the AMP-dependent protein kinase in vivo; Munday et al. 1988; Hardie et al. 1989; Haystead and Hardie 1989 in preparation). Evidence has also been obtained that insulin does indeed promote the overall dephosphorylation and activation of acetyl-CoA carboxylase in vivo. The enzyme was found to be both reduced in phosphate content and activated when isolated from lactating mammary gland, or acini cells and also rat liver, from animals which has been refed after 24 h starvation, or treated with insulin after streptozotocin treatment (Munday and Hardie 1986). It would be intriguing to carry out quantitative mapping of the phosphorylation sites on enzymes isolated from these tissues in order to determine any involvement of CKII in the activation process.

3 Summary and Future Work

Insulin, EGF, and PDGF work through common mechanisms to bring about the activation of glycogen synthase, CKII, and $PP\text{-}1_I$ in isolated cells. The activity of $PP\text{-}1_I$ appears to be acutely attenuated by the activity of CKII, implicating an important physiological role for this protein kinase. The relative activities of CKII and $PP\text{-}1_I$ may offer an explanation for the refractory response of glycogen synthase activation and then inhibition, following hormonal stimulation of 3T3 cells. In order to establish further the involvement of CKII activity in the regulation of glucose metabolism several questions remain to be answered:

1. Determine the mechanism by which insulin and growth factor receptor tyrosine kinase activation bring about the activation of CKII.
2. Correlate directly the activities of CKII and $PP\text{-}1_I$ in response to hormone stimulation with changes in the activity state of glycogen synthase by mapping of the phosphorylation sites of inhibitor 2 and glycogen synthase.
3. Ascertain the functional role of CKII phosphorylation in acetyl-CoA carboxylase and possibly other known CKII substrates in vivo.
4. Look for other examples of insulin and growth factor stimulation of CKII phosphorylation sites in other known substrates in vivo, such as glycogen synthase, inhibitor 2, and RII (regulatory subunit II of cyclic-A kinase).

References

Ackerman, P., Glover, C.V.C. & Osheroff, N. (1985) Phosphorylation of DNA topoisomerase II by casein kinase II: Modulation of eukaryotic topoisomerase II activity in vitro. Proc. Natl. Acad. Sci. USA 82:3164–3168

Ackerman, P., Glover C.V.C. & Osheroff, N. (1988) Phosphorylation of DNA topoisomerase II in vivo and in total homogenates of *Drosophila* cells: the role of casein kinase II. J. Biol. Chem. 263:12653–12660

Ballou, L.M. & Fischer, E.H. (1986) Phosphoprotein phosphatases. In: Boyer, P. & Krebs, E.G. (eds.) The enzymes, vol. 17. Academic Press, New York London Orland, pp. 311–361

Ballou, L.M., Brautigan, D.L. & Fischer, E.H. (1983) Subunit structure and activation of inactive phosphorylase phosphatase. Biochemistry 22:3393–3399

Brautigan, D.L., Picton, C. & Fischer, E.H. (1980) Phosphorylase phosphatase complex from skeletal muscle. Activation of one of two catalytic subunits by manganese ions. Biochemistry 19:5787–5794
Brautigan, D.L., Ballou, L.M. & Fischer, E.H. (1982) Biochemistry 21:1977–1982
Caudwell, F.B., Hiraga, C. & Cohen, P. (1986) Amino acid sequence of a region on the glycogen-binding subunit of protein phosphatase-1 phosphorylated by cyclic AMP-dependent protein kinase. FEBS Lett. 194:85–89
Chisholm, A.A.K. & Cohen, P. (1988) Identification of a third form of protein phosphatase 1 in rabbit skeletal muscle that is associated with myosin. Biochim. Biophys. Acta 968:392–400
Cohen, P. (1986) Muscle glycogen synthase. In: Boyer, P. & Krebs, E.G. (eds) The enzymes, vol. 17. Academic Press, New York London Orlando, pp. 461–495
Czech, M.P., Klarlund, J.K., Yagaloff, K.A., Bradford, A.P. & Lewis, R.E. (1988) Insulin receptor signaling. J. Biol. Chem. 263:11017–11020
Denton, R.M. (1986) Early events in insulin actions. In: Greengard, P. & Robinson, G.R. (eds) Advances in cyclic nucleotide and protein phosphorylation research, vol. 29. Raven, New York, pp. 293–341
DePaoli-Roach, A.A. (1984) Synergistic phosphorylation and activation of ATP-Mg-dependent phosphoprotein phosphatase by FA/GSK-3 and casein kinase II. J. Biol. Chem. 259:12144–12152
Erikson, E. & Maller, J.L. (1986) Purification and characterization of a protein kinase from Xenopus eggs highly specific for ribosomal protein S6. J. Biol. Chem. 261:350–355
Fiol, C.J., Mahrenholz, A.M., Wang, Y., Roeske, R.W. & Roach, P.J. (1987) Formation of protein kinase recognition sites by covalent modification of the substrate. J. Biol. Chem. 262:14042–14048
Giugni, T.D., Chen, K. & Cohen, S. (1988) Activation of a cytosolic serine protein kinase by epidermal growth factor. J. Biol. Chem. 263:18988–18995
Goris, J., Defreyn, G. & Merlevede, W. (1979) Resolution of the ATP-Mg-dependent phosphorylase phosphatase from liver into a two protein component system. FEBS Lett. 32:279–282
Hardie, D.G., Carling, D. & Sim, A.T.R. (1989) The AMP-activated protein kinase: a multisubstrate regulator of lipid metabolism. TIBS 14:20–23
Haystead, T.A.J. & Hardie, D.G. (1986) Evidence that activation of acetyl-CoA carboxylase by insulin in adipocytes is mediated by a low-M_r effector and not by increased phosphorylation. Biochem. J. 240:99–106
Haystead, T.A.J. & Hardie, D.G. (in preparation) A role for the AMP-activated protein kinase in the cyclic AMP-mediated inactivation of acetyl-CoA carboxylase in rat adipocytes
Haystead, T.A.J., Campbell, D.G. & Hardie, D.G. (1988) Analysis of sites phosphorylated on acetyl-CoA carboxylase in response to insulin in isolated adipocytes. Eur. J. Biochem. 175:347–354
Hemmings, B.A., Resink, T.J. & Cohen, P. (1982) Reconstitution of a Mg-ATP-dependent protein phosphatase and its activation through a phosphorylation mechanism. FEBS Lett. 150:319–324
Hiraga, A. & Cohen, P. (1986) Phosphorylation of the glycogen-binding subunit of protein phosphatase-1_G by cyclic-AMP-dependent protein kinase promotes translocation of the phosphatase from glycogen to cytosol in rabbit skeletal muscle. Eur. J. Biochem. 161:763–769
Holland, R. & Hardie, D.G. (1985) Both insulin and epidermal growth factor stimulate fatty acid synthesis and increase phosphorylation of acetyl-CoA carboxylase and ATP-citrate lyase in isolated hepatocytes. FEBS Lett. 181:308–312
Holmes, C.F.B., Campbell, D.G., Caudwell, B., Aitken, A. & Cohen, P. (1986a) Eur. J. Biochem. 155:173–182
Holmes, C.F.B., Kuret, J. Chisholm, A.A.K. & Cohen, P. (1986b) Identification of the sites on rabbit skeletal muscle protein phosphatase inhibitor-2 phosphorylated by casein kinase II. Biochim. Biophys. Acta 870:408–416
Holmes, C.F.B., Tonks, N.K., Major, H. & Cohen, P. (1987) Analysis of the in vivo phosphorylation state of protein phosphatase inhibitor-2 from rabbit skeletal muscle by fast-atom bombardment mass spectrometry. Biochim. Biophys. Acta 929:208–219
Jarett, L., Kiechle, F., Macaulay, S.L., Parker, J.C. & Kelly, K.L. (1985) Intracellular mediators of insulin action. In: Czech, M.P. (ed.) Molecular basis of insulin action. Plenum, New York, pp. 183–198
Jenö, P., Jäggi, N., Luther, H., Siegmann, M. & Thomas, G. (1989) Purification and characterization of a 40 S ribosomal protein S6 kinase from vanadate-stimulated Swiss 3T3 cells. J. Biol. Chem. 264:1293–1297
Jurgensen, S., Shacter, E., Huang, C.Y. & Chock, P.B., Yang, S.-D., Vandenheede, J.R. & Merlevede, W. (1984) On the mechanism of activation of the ATP•Mg(II)-dependent phosphoprotein phosphatase by kinase FA. J. Biol. Chem. 259:5864–5870
Kelly, K.L., Merida, I., Wong, E.H.A., DiCenzo, D. & Mato, J.M. (1987) A phospho-oligosaccaride mimics the effect of insulin to inhibit isoproterenol-dependent phosphorylation of phospholipid methyltransferase in isolated adipocytes. J. Biol. Chem. 262:15285–15290

Klarlund, J.K. & Czech, M.P. (1988) Insulin-like growth factor I and insulin rapidly increase casein kinase II activity in BALB/c 3T3 fibroblasts. J. Biol. Chem. 263:15872–15875

Larner, J., Gastio, G., Cheng, K., DePaoli-Roach, A.A., Huang, L., Daggy, P. & Kellogg, J. (1979) Generation by insulin of a chemical mediator that controls protein phosphorylation and dephosphorylation. Science 206:1408–1410

Lawrence, Jr., J.C. & Hinken, J.F. (1983) Hormonal control of glycogen synthase in rat hemidiaphragms. J. Biol. Chem. 258:10710–10719

MacKintosh, C., Campbell, D.G., Hiraga, A. & Cohen, P. (1988) Phosphorylation of the glycogen-binding subunit of protein phosphatase-1_G in response to adrenalin. FEBS Lett. 234:189–194

Munday, M.R. & Hardie, D.G. (1986) The role of acetyl-CoA carboxylase phosphorylation in the control of mammary gland fatty acid synthesis during the starvation and re-feeding of lactating rats. Biochem. J. 237:85–91

Munday, M.R., Campbell, D.G., Carling D. & Hardie, D.G. (1988) Identification by amino acid sequencing of three major regulatory phosphorylation sites on rat acetyl-CoA carboxylase. Eur. J. Biochem. 175:331–338

Novak-Hofer, I. & Thomas, G. (1985) Epidermal growth factor-mediated activation of an S6 kinase in Swiss mouse 3T3 cells. J. Biol. Chem. 260:10314–10319

Palen, E. & Traugh, J.A. (1987) Phosphorylation of ribosomal protein S6 by cAMP-dependent protein kinase and mitogen-stimulated S6 kinase differentially alters translation of globin mRNA. J. Biol. Chem. 262:3518–3523

Parker, P.J., Caudwell, F.B. & Cohen (1983) Glycogen synthase from rabbit skeletal muscle. Effect of insulin on the state of phosphorylation of the seven phosphoserine residues in vivo. Eur. J. Biochem., 130:227–234

Pelech, S.L. Tinker, D.A., Chan, C.P. & Krebs, E.G. (1987) Role of protein phosphorylation in growth factor signal transduction. In: Raizada, M.M., Phillips, M.I. & LeRoith, D. (eds.) Insulin, insulin-like growth factors, and their receptors in the central nervous system. Plenum, New York, pp 27–46

Ray, L.B. & Sturgill, T.W. (1987) Rapid stimulation by insulin of a serine/threonine kinase in 3T3-L1 adipocytes that phosphorylates microtubule-associated protein 2 in vitro. Proc. Natl. Acad. Sci. USA 84:1502–1506

Ray, L.B. & Sturgill, T.W. (1988a) Characterization of insulin stimulation microtubule-associated protein kinase, J. Biol. Chem. 263:12721–12727

Ray, L.B. & Sturgill, T.W. (1988b) Insulin stimulation microtubule-associated protein kinase is phosphorylated on tyrosine and threonine in vivo. Proc. Natl. Acad. Sci. USA 85:3753–3757

Reed, L.J. & Yeamen, S.J. (1986) Pyruvate dehydrogenase. In: Boyer, P. & Krebs, E.G. (eds.) The enzymes, vol. 17. Academic Press, New York London Orlando, pp. 77–95

Roach, P.J. (1986) Liver glycogen synthase. In: Boyer, P. & Krebs, E.G. (eds.) The enzymes, vol. 17. Academic Press, New York London Orlando, pp. 499–539

Rosen, O.M. (1987) Aftern insulin binds. Science 237:1452–1458

Ross, R., Nist, C., Kariya, B., Rivest, M.J., Raines, E. & Callis (1978) Physiological quiescence in plasma-derived growth factor on cell growth in culture. J. Cell. Physiol. 97:497–508

Saltiel, A.R., Fox, J.A., Sherline, P. & Cuatrecases, P. (1986) Insulin-stimulated hydrolysis of a novel glycolipid generates modulators of cAMP phosphodiesterase. Science 233:967–972

Sommercorn, J., McNall, S.J., Fischer, E.H. & Krebs, E.G. (1987) Phosphorylation of acetyl-CoA carboxylase by casein kinase II enhances the rate of dephosphorylation of the cAMP-dependent protein kinase site. Fed. Proc. 46:2003, Abst 452

Stewart, A.A., Hemmings, B.A., Cohen, P., Goris, J. & Merlevede, W. (1981) The Mg-ATP, dependent protein phosphatase and protein phosphatase 1 have identical substrate specificities. Eur. J. Biochem. 205:196–205

Strålfors, P., Fredrikson, G., Olsson, H. & Belfrage, P. (1984) Reversible phosphorylation of hormone-sensitive lipase in the hormonal control of adipose tissue lipolysis. Hormone Cell. Reg. 8:153–162

Strålfors, P., Hiraga, A. & Cohen, P. (1985) The protein phosphatases involved in cellular regulation. Eur. J. Biochem. 149:295–303

Sturgill, T.W., Ray B.L., Erikson, E. & Maller, J.L. (1988) Insulin stimulates MAP-2 kinase phosphorylates and activates ribosomal protein S6 kinase II. Nature (London) 334:715–718

Villa-Moruzzi, E., Ballou, L.M. & Fischer, E.H. (1984) Phosphorylase phosphatase. J. Biol. Chem 259:5857–5863

Witters, L.A., Tipper, J.P. & Bacon, G.W. (1983) Stimulation of site-specific phosphorylation of acetyl coenzyme A carboxylase by insulin and epinephrine. J. Biol. Chem. 258:5643–5648

Witters, L.A., Watts, T.D., Daniels, D.L. & Evans, J.L. (1988) Insulin stimulates the dephosphorylation and activation of acetyl-CoA carboxylase. Proc. Natl. Acad Sci. USA 85:5473–5477

Yang, S.-D., Chou, C., Huang, M., Song, J.-S. & Chen, H.-C. Epidermal growth factor induces activation of protein kinase FA and ATP•Mg-dependent protein phosphatase in A431 cells. J. Biol. Chem 264:5407–5411

Yang, S.-D., Ho, L.-T. & Fung, T.-J. (1988) Insulin induces activation and translocation of protein kinase FA (a multifunctional protein phosphatase activator) in human platelet. Biochem. Biophys. Res. Commun. 151:61–69

Yang, S.-D., Vandenheede, J.R., Goris, J. & Merlevede, W. (1980) ATP•Mg-dependent protein phosphatase from rabbit skeletal muscle. J. Biol. Chem 255:11759–11767

Part 2

Genes Coding for G Proteins in Mammalian and Yeast Cells

Y. Kaziro[1]

1 Introduction

GTP-binding proteins are classified largely into two groups, i.e., heterotrimeric GTP-binding proteins which are referred to as G proteins, and low-molecular-weight monomeric GTP-binding proteins (LMG) including *ras, rap, rho,* and *rab* proteins. The basic mechanism of the reaction catalyzed by these proteins appears to be analogous to that proposed for translational elongation factors (Kaziro 1978). The GTP-bound form is an active conformation which activates the transmission of signals, and the hydrolysis of bound GTP to GDP is required to shift the conformation to an inactive form, i.e., to shut off the signal transduction.

G proteins are involved in a variety of transmembrane signaling systems as transducers (for reviews, see Stryer and Bourne 1986; Gilman 1987). Two G proteins, Gs and Gi, are involved in hormonal stimulation and inhibition, respectively, of adenylate cyclase, whereas Go (another G protein), which is present predominantly in brain tissues, may be involved in neuronal responses. Two transducins, Gt1 and Gt2, which are present in retinal rods and cones, respectively, regulate cGMP phosphodiesterase activity and mediate visual signal transduction. There is evidence suggesting the presence of additional G proteins, which may be involved in the activation of phospholiapase C and phospholipase A_2, as well as the gating of K^+ and Ca^{2+} channels.

In this chapter, I would like to give a brief review of the structure of cDNAs for various G protein α-subunits (Gαs), the organization of human genes for Gαs, and the occurrence of two Gα genes in *Saccharomyces cerevisiae.*

2 Results and Discussion

2.1 Isolation of cDNA Clones for G Protein α-Subunits from Mammalian Cells

Recently, much effort has been focused on the cloning of cDNAs coding for various G protein α-subunits. These studies have revealed that there are at least three subtypes of Giα whose structures are closely related but distinct (reviewed in Kaziro et al. 1988). Transducins (Gt) also have two subtypes, Gt1α and Gt2α, which are expressed in rods and cones, respectively (Lerea et al. 1986). On the other hand, four different Gsα cDNAs are generated by alternative splicing as described below.

[1]Institute of Medical Science, University of Tokyo, 4-6-1, Shirokanedai, Minato-ku, Tokyo 108, Japan

40. Colloquium Mosbach 1989
Molecular Mechanisms of Hormone Action

Studies on the cDNA cloning of G protein α-subunits from rat C6 glioma cells revealed the structure of Gsα, Gi2α, Gi3α, and Goα, consisting of 394, 355, 354, and 354 amino acid residues, respectively, with molecular weights of 45 663, 40 499, 40 522, and 40 068 (Itoh et al. 1986, 1988). Gi1α, which is the most abundant among Giα subtypes in brain tissues, was not expressed in C6 glioma cells. However, Gi1α cDNA clones were obtained from rat olfactory epithelium (Jones and Reed 1987) and brain (Itoh et al. unpublished). We have recently isolated a new Gα clone (designated Gxα) which is apparently insensitive to pertussis toxin (Matsuoka et al. 1988). The Cys residue at the fourth position from the C-terminus, which is common to all pertussis toxin-sensitive Gαs, is replaced by Ile in Gxα. Human Gzα cDNA, isolated independently by Fong et al. (1988) from retina, may be the counterpart of Gxα. Rat Gi1α and Gxα code for 354 and 355 amino acid residues, respectively, with molecular weights of 40 345 and 40 879. The amino acid sequences of rat Gsα, Gi2α, Gi3α, Gi1α, Goα, and Gxα deduced from the nucleotide sequence are shown in Fig. 1.

Figure 2 shows a schematic representation of the structure of *E. coli* EF-Tu, G protein α-subunits (Gα), yeast RAS2 protein, and mammalian H-*ras* p21 protein. A remarkable homology was found in two regions, designated as P and G sites of all GTP-binding proteins. In EF-Tu, earlier biochemical studies indicated that the region around Cys-137 of EF-Tu (G site) is responsible for interaction with the guanine nucleotide (Kaziro 1978), and later the four residues Asn-Lys-Cys-Asp were found to be situated close to the guanine ring by X-ray analysis (Jurnak 1985). On the other hand, it has been shown that the mutation of amino acid residue 12 of p21 from Gly to Val decreases GTPase activity and increases transforming activity. Sequence homology to this region (P site) was found in all GTP binding proteins. The consensus sequence, GXXXGK, was located close to phosphoryl residues of bound guanine nucleotide (Jurnak 1985; deVos et al. 1988). The G′ site is a unique sequence which is

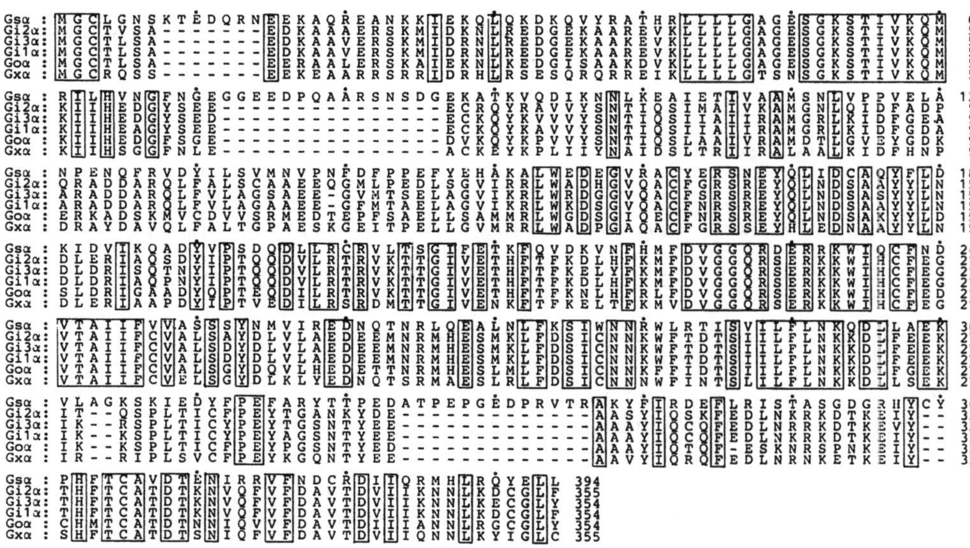

Fig. 1. Deduced amino acid sequences of rat Gsα, Gi2α, Gi3α, Gi1α, Goα, and Gxα. (Kaziro et al. 1988)

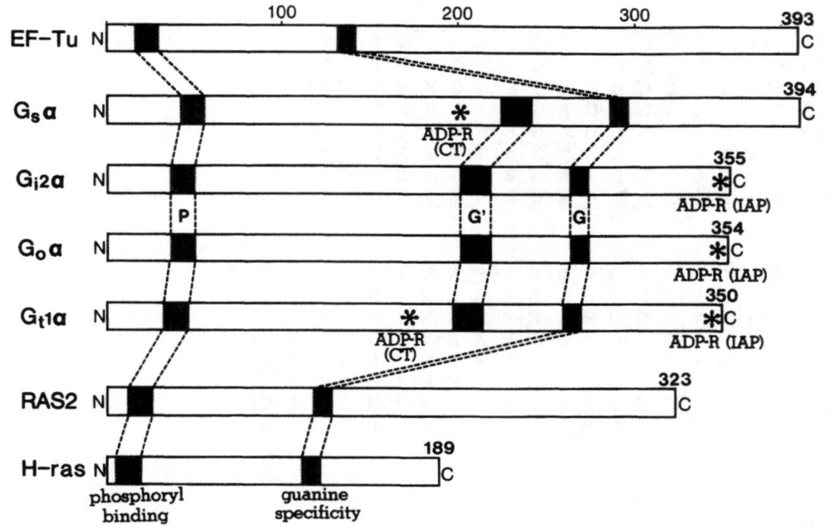

Fig. 2. Schematic representation of structures of EF-Tu, Gsα, Gi2α, Goα, Gtlα, RAS2, and H-ras p21

highly conserved in all G proteins but is not as remarkable in other GTP binding proteins except for *ARF* proteins (Sewell and Kahn 1988). Only the amino terminal sequence of the G′ site, Asp-Xaa-Xaa-Gly, is conserved in all GTP-binding proteins. The DXXG sequence is found in p21 at residues 57–60, and in *E. coli* EF-Tu at residues 80–83. The deduced amino acid sequences of Gαs in the P, G′, and G sites are shown in Fig. 3A-C, respectively.

It must be noted that the predicted amino sequence of a new Gxα was different from other Gα proteins at the P site. As shown in Fig. 3A, three amino acid residues in consensus to GTP-hydrolysis site of Gxα (Thr-Ser-Asn, at position 41–43) are different from the corresponding residues in other Gα proteins (Ala-Gly-Glu). It remains to be seen whether the kinetics of the Gxα-mediated signal transduction are different from the other systems due to the replacement of Gly (which corresponds to Gly-12 of p21) to Ser.

2.2 *Isolation of Human Gα Genes*

We have screened human genomic libraries with above rat cDNA clones and isolated human genes coding for Gsα, Gi1α, Gi2α, Gi3α, and Goα. So far, we have determined the gene organization and nucleotide sequences of total exons of Gsα, Gi2α, and Gi3α, and obtained the partial sequences (exons 1, 2, and 3) of Gi1α (Kozasa et al. 1988; and Itoh et al. 1988). The human Goα gene is a huge gene spanning at least 90 kb, however, the organization of exons is completely identical to those of Giα subfamily (Tsukamoto et al. unpublished).

(A)

```
  Gsα:    42-57: R L L L L G A G E S G K S T I V
 Gi1α:    35-50: K L L L L G A G E S G K S T I V
 Gi2α:    35-50: K L L L L G A G E S G K S T I V
 Gi3α:    35-50: K L L L L G A G E S G K S T I V
  Goα:    35-50: K L L L L G A G E S G K S T I V
  Gxα:    35-50: K L L L L G T S N S G K S T I V
 GP1α:    43-58: K L L L L G A G E S G K S T V L
 GP2α:  125-140: K V L L L G A G E S G K S T V L
```

(B)

```
  Gsα:  223-240: D V G G Q R D E R R K W I Q C F N D
 Gi1α:  200-217: D V G G Q R S E R K K W I H C F E G
 Gi2α:  201-218: D V G G Q R S E R K K W I H C F E G
 Gi3α:  200-217: D V G G Q R S E R K K W I H C F E G
  Goα:  201-218: D V G G Q R S E R K K W I H C F E D
  Gxα:  201-218: D V G G Q R S E R K K W I H C F E G
 GP1α:  319-336: D A G G Q R S E R K K W I H C F E G
 GP2α:  296-313: D V G G Q R S E R K K W I H C F D N
```

(C)

```
  Gsα:  285-296: I S V I L F L N K Q D L
 Gi1α:  262-273: T S I I L F L N K K D L
 Gi2α:  263-274: T S I I L F L N K K D L
 Gi3α:  262-273: T S I I L F L N K K D L
  Goα:  263-274: T S I I L F L N K K D L
  Gxα:  263-274: T S L I L F L N K K D L
 GP1α:  381-392: T P F I L F L N K I D L
 GP2α:  358-369: T S V V L F L N K I D L
```

Fig. 3A-C. Conserved sequences of G proteins. **A** Sequences of P site; **B** G′ site; **C** G site

2.3 Structure of the Human Gsα Gene and Generation of Four Gsα cDNAs by Alternative Splicing

The human Gsα gene isolated by T. Kozasa et al. (1988) contained 13 exons and 12 introns and spanned about 20 kb of genomic DNA. It has been known that there are two species of Gsα protein with different molecular masses (45 and 52 kDa) (Northup et al. 1980). Recently, Bray et al. (1986) isolated four different Gsα cDNAs (Gsα-1 to 4) from human brain and characterized the partial structure. Gsα-1 and Gsα-3 are identical except that Gsα-3 lacks a single stretch of 45 nucleotides. Gsα-2 and Gsα-4 have three additional nucleotides (CAG) to Gsα-1 and Gsα-3 3′ to the above 45 nucleotides. Robishaw et al. (1986a,b) isolated also two Gsα cDNAs from bovine adrenal gland that correspond to Gsα-1 and Gsα-4, and showed that these two cDNAs generated a 52- and 45-kDa protein when expressed in COS-m6 cells. Mattera et al. (1986) also isolated two Gsα cDNAs from human liver that correspond to Gsα-1 and Gsα-4.

Comparison of the four types of human Gsα cDNAs reported by Bray et al. (1986) with the sequence of the human Gsα gene of Kozasa et al. (1988) suggests that four types of Gsα mRNAs may be generated from a single Gsα gene by alternative splicing as shown in Fig. 4. Gsα-1 has a sequence identical to exons 2, 3, and 4, whereas Gsα-3 lacks exon 3. Gsα-2 and Gsα-4 have three additional nucleotides (CAG) to Gsα-1 and Gsα-3, respectively, at the 5′ end of exon 4. This CAG sequence is found in the genomic

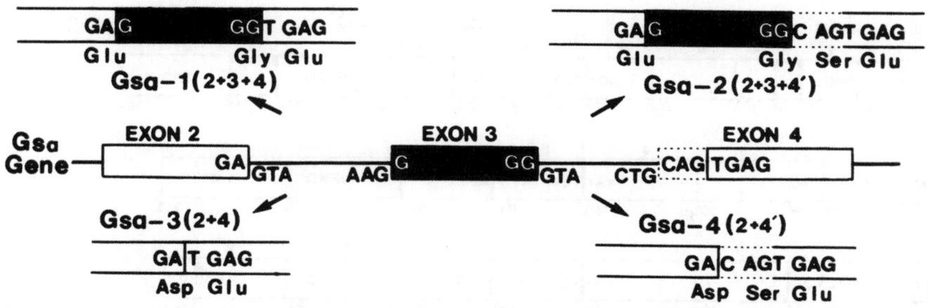

Fig. 4. Generation of four different Gsα mRNAs by alternative splicing. The Gsα gene is shown in the *center*. Gsα mRNAs are indicated by Gsα-1, -2, -3, and -4. For details, see Kozasa et al. (1988)

sequence of the 3′ splice site of intron 3. One additional serine residue, which is inserted upstream of exon 4 in Gsα-2 and Gsα-4, may be the potential site for phosphorylation by protein kinase C, and the alternative use of these splice sites may confer Gsα proteins with differential regulatory properties.

2.4 Human Genes for Giα Subtypes

The coding region of the human Gi2α and Gi3α genes splits into 8 exons and 7 introns (Itoh et al. 1988). There is an additional exon (exon 9) in the 3′ non-coding region of Gi2α and Gi3α. For human Gi1α, we only have the sequences of exons 1 to 3 at present. Remarkably, the positions of the splice junctions on the sequence of cDNA for Gi2α and Gi3α were completely identical (Fig. 5), although the length of introns are different (Itoh et al. 1988). The same splice sites are also conserved in the partial sequence (exons 1, 2, and 3) of the human Gi1α gene as well as in the human Goα gene (T. Tsukamoto unpublished). From the Southern blot analysis, it appears that each of the three Giα genes occurs as a single copy per haploid human genome.

2.5 Organization of Human Gα Genes

The exon-intron organization of the Gsα, Gi2α, Gi3α, and Goα genes was compared with the predicted functional domain structure of proteins (Fig. 5). The NH_2-terminal domain encoded by exon 1 is hydrophilic and contains the site for limited tryptic digestions. Although this region may be involved in interaction with βγ subunits, its precise function has not yet been elucidated. Exon 2 encodes a short length region (24 and 14 amino acid residues, respectively, for Gsα and Giαs), which is the most conserved among all Gα proteins and responsible for GTP hydrolysis. Exon 3 of Gsα is the one which is unique to Gsα. This exon is lost by alternative splicing in some of the subtypes of Gsα. The domain encoded by exons 4 to 6 of Gsα and 3 to 4 of Giα is structurally divergent. Exon 8 of Gsα contains Arg-201, which is ADP-ribosylated in the presence of cholera toxin (Van Dop et al. 1984). ADP-ribosylation of Gsα by cholera toxin causes a decrease of affinity for βγ subunits (Kahn and Gilman 1984). Arg-179 in exon 5 of Gi2α corresponds to this arginine residue. The domain encoded by exons 9 to 11 of Gsα, and 6 to 7 of Giα is strongly conserved among all Gα proteins.

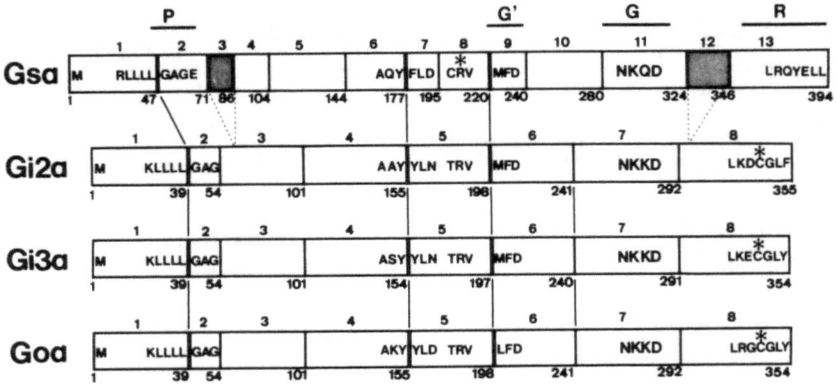

Fig. 5. Organization of the exons of mammalian G protein α-subunits

This domain is involved in the formation of a core structure for GTP binding together with that coded by exon 2 (Masters et al. 1986). The sequence, Asn-Lys-Xaa-Asp, in consensus to all guanine nucleotide binding proteins, occurs in exon 11 of Gsα and exon 7 of Giα. The conserved Asp-223 in exon 9 of Gsα and Asp-201 in exon 6 of Gi2α may form a salt bridge to Mg^{2+}, which is linked to the β-phosphoryl group of GDP (Jurnak 1985). The exchange of GDP to GTP may result in displacement of the surrounding region residues 230–238 in exon 9 of Gsα. A nonhydrolyzable GTP analog, but not GDP, prevents tryptic cleavage at Lys-210 in Goα or Lys-205 in Gtlα (Hurley et al. 1984).

Exon 12 of Gsα is unique to Gsα, and exon 13 of Gsα and exon 8 of Giα encode the COOH terminus region. The domain may be involved in interaction with a receptor, since the Cys residue which is ADP-ribosylated by pertussis toxin is present in this region of Giα, and also the structure of this region is heterogeneous. In Gxα, the Cys residue is replaced by Ile, indicating that Gxα is probably refractory to modification by pertussis toxin. Gsα, which is also resistant to pertussis toxin, possessed Tyr instead of Cys in this position. It was shown that the replacement of Arg to Pro at –6 position of Gsα gives rise to a mutant protein which is uncoupled with β-adrenergic receptor in S49 cells (Sullivan et al. 1987). More recently, Masters et al. (1988) have shown, using the chimeric Gi/Gs construct, that the carboxy terminal domain of Gsα contains the structure specifying interactions with the effector enzyme, adenylate cyclase, as well as with the hormone receptor. Further studies, including site-directed mutagenesis and construction of chimeric genes, may throw more light on the structure-function relationship of Gα proteins.

Comparison of the exon organization of the Giα subfamily and Goα with that of Gsα indicated that some of the exon junctions were conserved between the Giα subfamily and Gsα. Thus, 3 of 12 splice sites of the human Gsα gene are shared with the human Giα genes, and exon 1 and exons 7 and 8 of Gsα correspond to exon 1 and 5 of Giα, respectively.

2.6 Conservation of the Primary Structure of Each Gα Among Mammalian Species

Table 1 shows that, in addition to the remarkable homologies of the overall structure, there is a strong conservation of the amino acid sequence in each subtype of G protein α-subunit. The amino acid sequence of Gsα is strongly conserved between human and rat; only 1 of 394 amino acids being different. The sequence of Gi1α is completely identical between bovine and human. For Gi2α, Gi3α, Gxα, and Goα, over 98% identity of amino acid sequences is maintained among different mammalian species.

The strong conservation of the amino acid sequence of each G protein α-subunit among distant mammalian species may reflect the presence of evolutionary pressure to maintain the specific physiological function of each G protein gene product.

Table 1. Conservation of G protein α-subunit sequences among different mammalian species

Species[a]			Amino acid sequences	Nucleotide sequences
rGsα	vs	hGsα	393/394 (99.7%)	1128/1182 (95.4%)
bGi1α	vs	hGi1α	354/354 (100%)	998/1062 (94.0%)
rGi2α	vs	hGi2α	350/355 (98.6%)	985/1065 (92.4%)
rGi3α	vs	hGi3α	349/354 (98.6%)	981/1062 (92.4%)
rGoα	vs	bGoα	348/354 (98.3%)	992/1062 (93.4%)
rGxα	vs	hGxα	349/355 (98.3%)	977/1065 (91.7%)

[a] h, Human; r, rat; b, bovine.

2.7 G Proteins from Saccharomyces cerevisiae

A family of GTP-binding protein, the *ras* family, is widely distributed among eukaryotes (see Barbacid 1987, for a review) including the yeast *Saccharomyces cerevisiae* (DeFeo-Jones et al. 1983; Powers et al. 1984) and *Schizosaccharomyces pombe* (Fukui and Kaziro 1985). It has been suggested that the *RAS2* gene in *S. cerevisiae* is involved in the activation of adenylate cyclase (Toda et al. 1985; Broek et al. 1985), and mimics the role of mammalian Gs.

However, in view of the strong conservation of the amino acid sequences of each G protein species among different organisms (see Table 1), we speculated that G protein may also occur in yeast. We have searched for G protein, homologous gene in yeast and isolated two genes *GPA1* (Nakafuku et al. 1987), and *GPA2* (Nakafuku et al. 1988) from *S. cerevisiae*, which are homologous to cDNAs for mammalian G protein α-subunits.

GPA1 and *GPA2* code for sequences of 472 and 449 amino acid residues, respectively, with calculated Mr's of 54 075 and 50 516. When aligned with the α-subunit of mammalian G proteins to obtain maximal homology, GP1α (*GPA1*-encoded protein) and GP2α (*GPA2*-encoded protein) were found to contain the stretches of 110 and 83 additional amino acid residues, respectively, near the NH_2 terminus (Fig. 6).

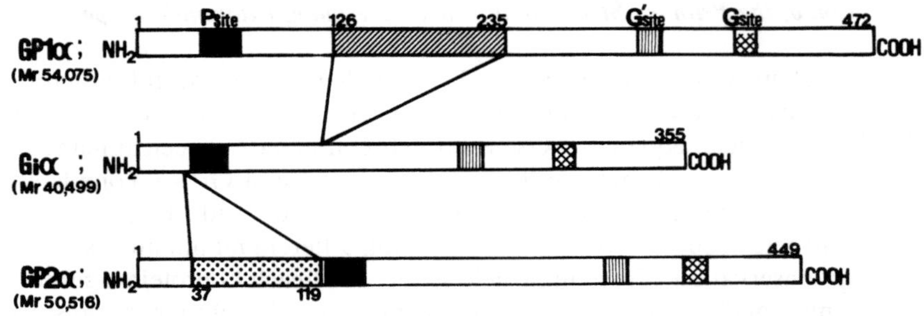

Fig. 6. Schematic representation of the structure of yeast GP1α, yeast GP2α, and mammalian Gi2α

2.8 Comparison of the Amino Acid Sequences of Yeast GP1α and GP2α with Those of Rat Brain Giα and Goα

The deduced amino acid sequence of yeast GP1α and GP2α is highly homologous to those of rat brain Giα and Goα. The homology is most remarkable in the region of GTP hydrolysis (P site, see Fig. 3A) (amino acid residues 43–58 of GP1α, 125–140 of GP2α, and 35–50 of Gi2α). As shown in Fig. 3C, the region responsible for GTP binding (G site) (amino acid residues 381–392 of GP1α, 358–369 of GP2α, and 263–274 of Gi2α) was also highly homologous. Another region of homology (G′ site, see Fig. 3B) was found in amino acid residues 319–336 of GP1α, 296–313 of GP2α, and 201–218 of Gi2α where a sequence of 14 contiguous amino acids were completely identical in yeast GP1α, GP2α, and rat Gi2α.

The overall homology in nucleotide and amino acid sequences of yeast GP1α, yeast GP2α, rat Gi2α, and rat Gsα is remarkable. Disregarding the unique sequences present in GP1α (residues 126–235) and GP2α (residues 37–119), the proteins are 60% homologous if conservative amino acid substitutions are considered to be homologous. The homology is smaller than that between rat Gi2α and Goα (85%) but is comparable to that between rat Gi2α and Gsα (60%).

As is described elsewhere in detail (Miyajima et al. 1987; Dietzel and Kurjan 1987), *GPA1* is a haploid-specific gene and involved in the mating factor signal transduction. On the other hand, *GPA2* is expressed both in haploid and diploid cells, and may be involved in the regulation of cAMP levels in *S. cerevisiae* (Nakafuku et al. 1988). The demonstration of the occurrence of G proteins in yeast may open the way for a detailed genetic analysis of the function of G proteins in eukaryotic cells.

References

Barbacid, M. (1987) *ras* Genes. Annu. Rev. Biochem. 56:779–827

Bray, P., Carter, A., Simons, C., Guo, V., Puckett, C., Kamholz, J., Spiegel, A., & Nirenberg, M. (1986) Human cDNA clones for four species of Gαs signal transduction protein. Proc. Natl. Acad. Sci. USA 83:8893–8897

Broek, D., Samiy, N., Fasano, O., Fujiyama, A., Tamanoi, F., Northup, J., & Wigler, M. (1985) Differential activation of yeast adenylate cyclase by wild-type and mutant *RAS* proteins. Cell 41:763–769

DeFeo-Jones, D., Scolnick, E.M., Koller, R., & Dhar, R. (1983) *ras*-Related gene sequences identified and isolated from *S. cerevisiae*. Nature (London) 306:707–709

DeVos, A.M., Tong, L., Milburn, M.V., Matias, P.M., Jankark, J., Noguchi, S., Nishimura, S., Miura, K., Ohtsuka, E., Kim, S.-H. (1988) Three-dimensional structure of an oncogene protein: catalytic domain of human c-H-*ras* p21. Science 239:888–893

Dietzel, D. & Kurjan, J. (1987) The yeast *SCG1* gene: a Gα-like protein implicated in the *a*- and α-factor response pathway. Cell 50:1001–1010

Fong, H.K.W., Yoshimoto, K.K., Eversole-Cire, P., & Simon, M.I. (1988) Identification of a GTP-binding protein α subunit that lacks an apparent ADP-ribosylation site for pertussis toxin. Proc. Natl. Acad. Sci. USA 85:3066–3070

Fukui, Y. & Kaziro, Y. (1985) Molecular cloning and sequence analysis of a *ras* gene from *Schizossacharomyces pombe*. EMBO J. 4:687–691

Gilman, A.G. (1987) G proteins: transducers of receptor-generated signals. Annu. Rev. Biochem. 56:615–649

Hurley, J.B., Simon, M.I., Teplow, D.B., Robishaw, J.D., & Gilman, A.G. (1984) Homologies between signal transducing G proteins and *ras* gene products. Science 226:860–862

Itoh, H., Kozasa, T., Nagata, S., Nakamura, S., Katada, T., Ui, M., Iwai, S., Ohtsuka, E., Kawasaki, H., Suzuki, K., & Kaziro, Y. (1986) Molecular cloning and sequence determination of cDNAs for α subunit of the guanine nucleotide-binding proteins Gs, Gi, and Go from rat brain. Proc. Natl. Acad. Sci. USA 83:3776–3780

Itoh, H., Toyama, R., Kozasa, T., Tsukamoto, T., Matsuoka, M., & Kaziro, Y. (1988) Presence of three distinct molecular species of Gi protein α subunit. J. Biol. Chem. 263:6656–6664

Jones, D.T. & Reed, R.R. (1987) Molecular cloning of five GTP-binding protein cDNA species from rat olfactory neuroepithelium. J. Biol. Chem. 262:14241–14249

Jurnak, F. (1985) Structure of the GDP domain of EF-Tu and location of the amino acids homologous to *ras* oncogene proteins. Science 230:32–36

Kahn, R.A. & Gilman, A.G. (1984) ADP-ribosylation of Gs promoters and the dissociation of its α and β subunits. J. Biol. Chem 259:6235–6240

Kaziro, Y. (1978) The role of guanosine 5′-triphosphate in polypeptide chain elongation. Biochim. Biophys. Acta 505:95–127

Kaziro, Y., Itoh, H., Kozasa, T., Tsukamoto, T., Matsuoka, M., Nakafuku, M., Obara, T., Takagi, T., & Hernandez, R. (1988) Structure of the genes coding for G protein α subunits from mammalian and yeast cells. Cold Spring Harbor Symp. Quant. Biol. 53:209–220

Kozasa, T., Itoh, H., Tsukamoto, T., & Kaziro, Y. (1988) Isolation and characterization of human Gsα gene. Proc. Natl. Acad. Sci. USA 85:2081–2085

Lerea, C.L., Somers, D.E., Hurley, J.B., Klock, I.B., & Bunt-Milam, A.H. (1986) Identification of specific transducin α subunits in retinal rod and cone photoreceptors. Science 324:77–80

Masters, S.B., Stroud, R.M., & Bourne, H.R. (1986) Family of G protein α chains: amphipathic analysis and predicted structure of functional domains. Protein Eng. 1:47–54.

Masters, S.B., Sullivan, K.A., Miller, R.T., Beiderman, B., Lopez, N.G., Ramachandran, J., & Bourne, H.R. (1988) Carboxyl terminal domain of Gsα specifies coupling of receptors to stimulation of adenylyl cyclase. Science 241:448–451

Matsuoka, M., Itoh, H., Kozasa, T., & Kaziro, Y. (1988) Sequence analysis of cDNA and genomic DNA for a putative pertussis toxin-insensitive guanine nucleotide-binding regulatory protein α subunit. Proc. Natl. Acad. Sci. USA 85:5384–5388

Mattera, R., Codina, J., Crozat, A., Kidd, V., Woo, S.L.C., & Birnbaumer, L. (1986) Identification by molecular cloning of two forms of the α-subunit of the human liver stimulatory (Gs) regulatory component of adenylate cyclase. FEBS Lett. 206:36–41

Miyajima, I., Nakafuku, M., Nakayama, N., Brenner, C., Miyajima, A., Kaibuchi, K., Arai, K., Kaziro, Y., & Matsumoto, K. (1987) Cell 50:1011–1019

Nakafuku, M., Itoh, H., Nakamura, S., & Kaziro, Y. (1987) Occurrence in *Saccharomyces cerevisiae* of a gene homologous to the cDNA coding for the α subunit of mammalian G proteins. Proc. Natl. Acad. Sci. USA 84:2140–2144

Nakafuku, M., Obara, T., Kaibuchi, K., Miyajima, I., Miyajima, A., Itoh, H., Nakamura, S., Arai, K., Matsumoto, K., & Kaziro, Y. (1988) Isolation of a second yeast *Saccharomyces cerevisiae* gene (*GPA2*) coding for guanine nucleotide-binding regulatory protein: studies on its structure and possible functions. Proc. Natl. Acad. Sci. USA 85:1374–1378

Northup, J.K., Sternweise, P.C., Smigel, M.D., Shleifer, L.S., Ross, E.M., and Gilman, A.G. (1980) Purification of the regulatory component of adenylate cyclase. Proc. Natl. Acad. Sci. USA 77:6516–6520

Powers, S., Kataoka, T., Fasano, O., Goldfarb, M., Strathern, J., Broach, J., & Wigler, M. (1984) Genes in *S. cerevisiae* encoding proteins with domains homologous to the mammalian *ras* proteins. Cell 36:607–612

Robishaw, J.D., Russell, D.W., Harris, B.A., Smigel, M.D., & Gilman, A.G. (1986a) Deduced primary structure of the α subunit of the GTP-binding stimulatory protein of adenylate cyclase. Proc. Natl. Acad. Sci. USA 83:1251–1255

Robishaw, J.D., Smigel, M.D., & Gilman, A.G. (1986b) Molecular basis for two forms of the G protein that stimulates adenylate cyclase. J. Biol. Chem. 261:9587–9590

Sewell, J.L. & Kahn, R.A. (1988) Sequences of the bovine and yeast-ADP-ribosylation factor and comparison to other GTP-binding proteins. Proc. Natl. Acad. Sci. USA 85:4620–4624

Stryer, L. & Bourne, H.R. (1986) G proteins: a family of signal transducers. Annu. Rev. Cell Biol. 2:391

Sullivan, K.A., Miller, R.T., Masters, S.B., Beiderman, B., Heideman, W., & Bourne, H.R. (1987) Identification of receptor contact site involved in receptor – G protein coupling. Nature (London) 330:758–760

Toda, T., Uno, I., Ishikawa, T., Powers, S., Kataoka, T., Broek, D., Cameron, S., Broach, J., Matsumoto, K., & Wigler, M. (1985) In yeast, *RAS* proteins are controlling elements of adenylate cyclase. Cell 40:27–36

Van Dop, C., Tsubokawa, M., Bourne, H.R., & Ramachandran, J. (1984) Amino acid sequences of retinal transducin at the site ADP-ribosylated by cholera toxin. J. Biol. Chem. 259:696–698

The *ras* Oncogene Protein

M.S. Marshall[1], M.D. Schaber[1], U.S. Vogel[1], W.S. Hill[1], A.S. Ng[1],
E.M. Scolnick[1], R.A.F. Dixon[1], I.S. Sigal[1], and J.B. Gibbs[1]

1 Introduction

To date, over 40 discrete genes (oncogenes and proto-oncogenes) have been shown capable of inducing cellular transformation and tumor development (Bishop 1987). Only a few of these oncogenes have actually been shown to have a significant role in human cancer. The proto-oncogenes *bcl-2*, c-*myc*, c-*abl*, c-*ets*, and c-*myb* are altered by chromosomal translocations in specific tumors. Amplification of the c-*erb*B, c-*myc*, L-*myc*, and N-*myc* proto-oncogenes occurs in a high frequency of other tumor types. Transfection of NIH3T3 cells with DNA isolated from human tumors frequently identifies the Ha-*ras*, Ki-*ras*, N-*ras*, *met*, *mel*, *trk*, *dbl*, and c-*raf*-1 oncogenes (Martin-Zanca et al. 1986; Ron et al. 1988).

The most commonly identified genetic lesions in human cancers involve the *ras* oncogenes. The normal *ras* genes encode homologous membrane-associated 21 kilodalton (kDa) proteins (for review, see Barbacid 1987). The *ras* proteins share similar properties with the 40 kDa G-proteins, including specific binding of GDP and GTP and GTP-hydrolytic activity (Gilman 1987). Oncogenic variants of the *ras*-encoded proteins differ from their cellular homologues by single amino acid substitutions at positions 12, 13, 61, and 63 which affect the intrinsic GTPase activity. The G-protein model suggests that the *ras* proteins function as membrane-bound signal transducers; active when bound to GTP and inactive when bound to GDP. By this model reduction of GTPase activity in oncogenic *ras* variants results in a greater proportion of the Ras*GTP complex and an increased signal. In most cases enhanced Ras activity results in cellular proliferation; however, the mechanism involved is not yet clear.

Insights into the biological role of the *ras* protein has been gained through studies of its structure and its biochemical properties as well as investigating its function in lower eukaryotes. The understanding obtained from this type of analysis relates to the mechanism of how Ras interacts with other proteins and how Ras activity is regulated. Additionally, a vertebrate protein called GTPase-activating protein (GAP) has been identified which may be involved in the *ras* pathway for growth control.

[1]Merck Sharp and Dohme Research Laboratories, West Point, PA 19486, USA
[2]Deceased

40. Colloquium Mosbach 1989
Molecular Mechanisms of Hormone Action

2 Results and Discussion

2.1 The Ras Effector Domain

The importance of the *ras* proteins in cellular functions is emphasized by the presence of *ras* homologues in many simple eukaryotic cell types. The yeast *Saccharomyces cerevisiae* possesses two genes highly homologous to Ha-*ras* (Gibbs and Marshall 1989). The primary distinction between Ha-*ras* and the yeast *RAS* proteins is the presence of unrelated 120 to 131 amino acid domains at the C-termini of the yeast *RAS* proteins. The proteins encoded by the *RAS1* and *RAS2* genes regulate adenylyl cyclase and are required for cell viability and cell cycle progression. The similarity between the human and yeast *ras* proteins is underscored by the ability of Ha-*ras* to promote the growth of yeast cells lacking both endogenous *RAS* genes.

Using yeast as a model system, mutations were introduced into the Ha-*ras* gene and regions of the protein identified which were involved in effector coupling (Sigal et al. 1986). Amino acid substitutions were found which reduced the biological activity of the *ras* protein without altering nucleotide binding or membrane localization. These effector mutations eliminated the ability of the oncogenic *ras* protein to either stimulate yeast cyclase or transform mammalian cells. These effector mutations were single amino acid changes at positions 32, 33, 35, 38, and 40 and were proposed to affect the region of Ras which interacts with other proteins. To test this hypothesis, a genetic screen was utilized to identify second-site mutations (intergenic mutations) which would promote the growth of *ras*-minus yeast in the presence of a *ras* effector mutant protein (Marshall et al. 1988). Theoretically, if the Ras 32–40 region was involved in protein to protein contact, then an appropriately altered target protein could recognize an effector mutant *ras* protein. A mutant yeast *RAS2* gene encoding a Ser for Thr substitution at position 42 (analogous to Ha-*ras* position 35) was used in this screen. One dominant mutation, *SSR2-1*, was identified which did conferred upon yeast a *ras* effector mutant-responsive phenotype. Upon testing the *SSR2-1* strain with other yeast *RAS* and Ha-*ras* genes, it was found that the mutant was also responsive to the [Asn 33]Ha-*ras* effector mutant protein. The *SSR2-1* mutation was genetically mapped and found to be allellic with the structural gene for adenylyl cyclase, *CYR1*. The *SSR2-1* gene was cloned, sequenced, and found to encode a Tyr for Asp substitution at position 1547 in the proposed regulatory domain of adenylyl cyclase. Biochemical analysis of the *SSR2-1* adenylyl cyclase confirmed that the enzyme was dependent upon Ras*GTP for activity and that its activity could be stimulated by *ras* effector mutant proteins. The isolation of a second-site suppressor of *ras* effector mutant proteins in the gene encoding a yeast *ras* target protein supports the model of the 32–40 region of Ras being involved in protein-protein interactions.

2.2 GAP-a Mammalian Protein that Interacts with ras p21

2.2.1 Purification and Characterization of GAP

Trahey and McCormick (1987) demonstrated the presence of a soluble factor in *Xenopus leavis* oocyte extracts which accelerated the rate of the intrinsic Ras GTP hydrolytic activity. This factor was active only on normal *ras* proteins, but not on oncogenic forms. The factor was named Ras GTPase-activating protein, or GAP. It was proposed that GAP might not be simply a regulator of *ras* p21 activity, but perhaps the long sought after mammalian Ras target protein. Just as interaction with the ribosome stimulates the GTPase activity of EF-Tu, Ras*GTP might stimulate GAP in an as yet unknown fashion, followed by inactivation by a GAP-catalyzed acceleration of the intrinsic GTPase rate of Ras.

GAP was purified to near homogeneity from the cytoplasm of bovine cerebra as a 125 kDa polypeptide and found to be identical in activity and molecular mass with the *Xenopus* protein (Gibbs et al. 1988). With purified GAP available, detailed characterization of the binding and enzymatic properties of the factor were performed (ibid; Sigal et al. 1989). Purified GAP stimulated the rate of GTP hydrolytic activity of Ras by more than 100-fold. This activity was specific for normal *ras* proteins but had no effect on the oncogenic [Val-12] and [Leu 61] *ras* variants or on the Ras-related protein YPT1. GAP also did not stimulate the GTPase activity of the biologically impaired effector mutant *ras* proteins. The ability of *ras* proteins to physically interact with GAP was determined by means of a sensitive kinetic competition assay. GAP interacts preferentially with the GTP but not the GDP complexes of both normal and oncogenic Ha-*ras* proteins. However, stimulation of intrinsic GTPase occurs only with the normal *ras* protein. With interest it was observed that the highly transforming [Leu 61]Ha-*ras* p21*GTP complex competed for GAP association at an IC_{50} value of only 2 μM compared to 110 μM for the normal Ha-*ras* p21*GTP complex. These results suggest that GAP interacts with *ras* proteins in the manner predicted for the interaction between mammalian Ras and its target.

2.2.2 *Cloning and Genetic Analysis of the GAP Gene*

Using peptide sequence obtained from purified bovine GAP, degenerated oligonucleotide probes were designed and used to clone the cDNA for bovine GAP (Vogel et al. 1988). The cloned sequence could encode a 116 kDa protein with a total amino acid composition similar to that of purified bovine GAP. Expression of the cloned sequence in *E. coli* resulted in the production of a 125 000 dalton protein with the same specific GTPase-stimulating activity as bovine GAP. The cDNAs for human and murine GAP were also cloned and were found to be nearly identical with bovine GAP with the exception of some minor divergence in their N-termini (Trahey et al. 1988; Vogel unpublished).

The amino acid sequence of bovine GAP was compared to the PIR protein data base and a translated Genbank sequence library. Although the statistical significance is questionable, sequence similarity was found between GAP and the regulatory domain of yeast adenylyl cyclase. More significant was the discovery that the N-terminal region of GAP contains two repeats of the SH2 domain (or B/C boxes) found in the regulatory region of the Src/nonreceptor tyrosine kinase family of proteins, phospholipase C type II, and the *crk* oncogene protein (see Fig. 1). The SH2 domain of Src and Fps has been shown to be regulatory in nature and may mediate association with other proteins (Sadowski et al. 1986; Wang and Parsons 1989). This domain might serve a similar function in GAP.

To identify the region of GAP responsible for Ras GTPase activation, a deletion analysis of the GAP coding region was performed in *E. coli* (Marshall et al. 1989). A

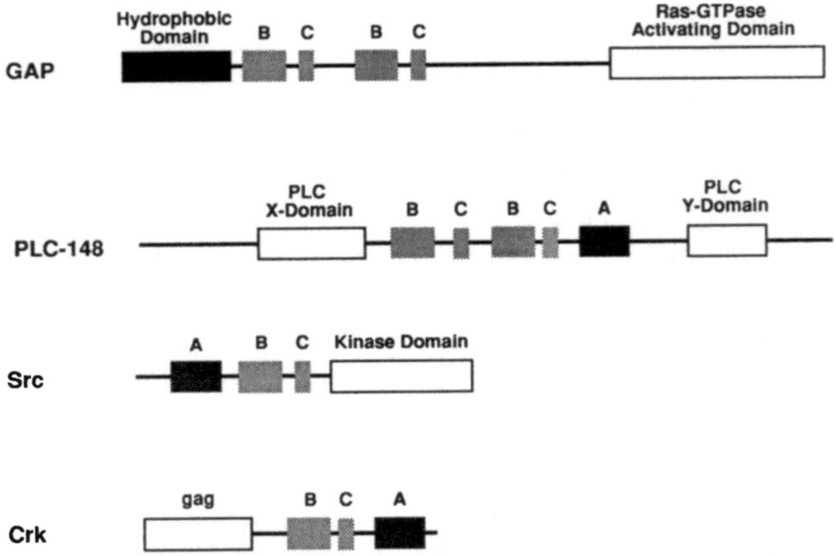

Fig. 1. Schematic diagram comparing the known structural domains of the Ras GTPase-activating protein with several growth regulatory proteins containing B/C box sequences

61 amino acid truncation of the GAP C-terminus resulted in an inactive, insoluble protein, while up to 701 amino acids could be truncated from the GAP N-terminus without inactivating its specific enzymatic activity (see Fig. 1). The smallest, active truncation of GAP ([702–1044]GAP) was purified from *E. coli* and characterized as to its specific activity and Ras-binding affinity. [702–1044]GAP was found to be identical to full-length purified bovine GAP in these two properties. These analyses have shown that GAP has a distinct-Ras-interactive domain at the C-terminus with a potential regulatory domain at the N-terminus.

2.2.3 GAP – Regulator or Effector?

Although the in vitro biochemical properties of GAP have been explored in great detail, the true function of GAP in the mammalian *ras* pathway remains unknown. The in vitro GTPase-activating properties of GAP clearly suggest that GAP could play a significant role in regulating the amount of GTP complexed to *ras* protein in the cell (Sigal 1988; McCormick 1989). Recently a gene, *IRA 1*, has been identified in the yeast *Saccharomyces cerevisiae*, which appears to promote the GDP-bound form of the *RAS2* protein (Tanaka et al. 1989). The protein encoded by the *IRA1* gene has significant amino acid sequence similarity with the GAP Ras-binding domain. GAP might be the vertebrate equivalent of the *IRA 1* protein and an upstream regulator of Ras activity (see Fig. 2). Alternatively, GAP may function as a downstream mediator of the Ras*GTP signal which incorporates an IRA1-like action as a feedback control mechanism (see Fig. 3). The *IRA 1* gene product differs from GAP in that it has no effect on Ha-*ras* expressed in yeast or added to yeast lysates. The *IRA 1* protein might interact

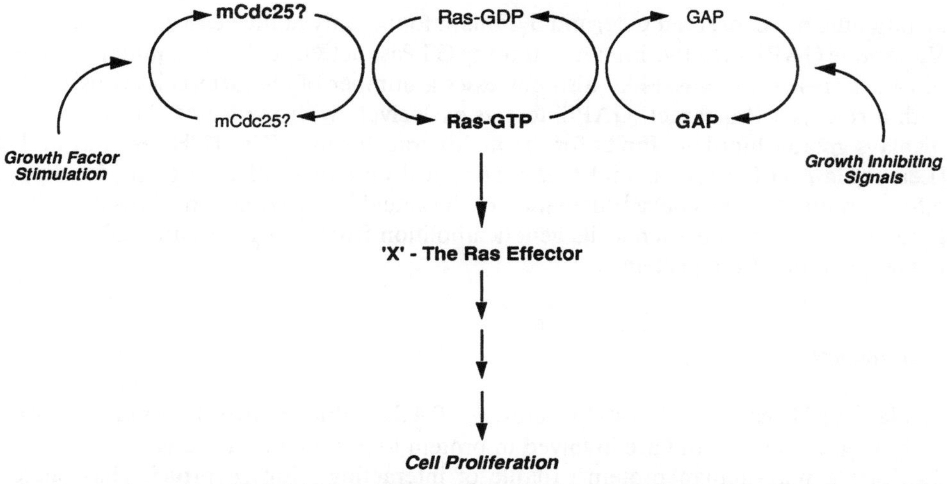

Fig. 2. A proposed model of GAP function based upon the *S. cerevisiae* Ras pathway. In this model GAP would function as a negative regulator of Ras function analogous to the *IRA* proteins of yeast

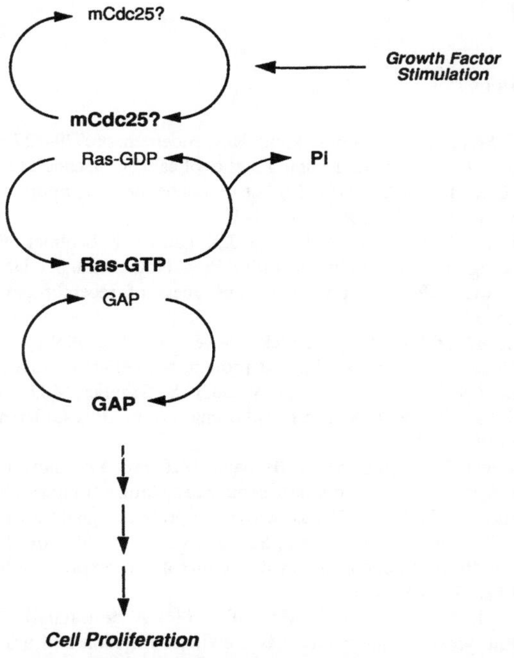

Fig. 3. A proposed model of GAP function based upon the interactions between *E. coli* EF-Tu and the ribosome. As suggested by this model GAP would be the target of Ras activity with the ability to down-regulate Ras following the interaction

through the nonconserved C-terminal domain found only in the yeast *RAS* proteins. Vertebrate GAP is effective in stimulating the GTPase activity of both mammalian and *S. cerevisiae ras* proteins. GAP also possesses a number of characteristics consistent with a role as a Ras target. GAP interacts exclusively with the Ras*GTP complex, displays greater binding affinity for the highly transforming [Val 12]Ha-*ras* p21 and [Leu 61]Ha-*ras* p21 proteins, and does not physically interact with biologically inactive effector mutant *ras* proteins. Elucidation of the actual in vivo function of GAP awaits biological experiments such as its genetic abolition from living cells and eukaryotic overexpression of the protein.

3 Summary

1. Ha-*ras* p21 residues 32 to 40 (*S. cerevisiae* RAS residues 39 to 47) are required for biological activity and are involved in protein to protein interactions.
2. GAP, a mammalian protein capable of interacting with *ras* protein, has been purified and the gene cloned.
3. In vitro enzymatic properties suggest that GAP acts as a regulator of Ras activity by accelerating the rate of GTP hydrolysis. The binding affinity of GAP for oncogenic and effector mutant *ras* proteins suggests that GAP could be the cellular target of *ras* p21.

References

Barbacid, M. (1987) ras Genes. Annu. Rev. Biochem. 56:779–827

Bishop, J.M. (1987) The molecular genetics of cancer. Science 235:305–311

Gibbs, J.B. & Marshall, M.S. (1989) The *ras* oncogene – an important regulatory element in lower eucaryotic organisms. Microbiol. Rev. 53:171–185

Gibbs, J.B., Schaber, M.D., Allard, W.J., Sigal, I.S. & Scolnick, E.M. (1988) Purification of ras GTPase activating protein from bovine brain. Proc. Natl. Acad. Sci. USA 85:5026–5030

Gilman, A.G. (1987) G proteins: transducers of receptor-generated signals. Annu. Rev. Biochem. 56:615–650

Marshall, M.S., Gibbs, J.B., Scolnick E.M. & Sigal, I.S. (1988) An adenylate cyclase from *Saccharomyces cerevisiae* that is stimulated by ras proteins with effector mutations. Mol. Cell. Biol. 8:52–61

Marshall, M.S., Hill, W.S., Ng, A.S., Vogel, U.S., Schaber, M.D., Scolnick, E.M., Dixon, R.A.F., Sigal, I.S. & Gibbs, J.B. (1989) A C-terminal domain of GAP is sufficient to stimulate *ras* p21 GTPase activity. EMBO J. 8:1105–1110

Martin-Zanca, D., Hughes, S.H. & Barbacid, M. (1986) A human oncogene formed by the fusion of truncated tropomyosin and protein kinase sequences, Nature (London) 319:743–748

McCormick, F. (1989) ras GTPase activating protein: signal transmitter and signal terminator. Cell 56:5–8

Ron, D., Tronick, S.R., Aaronson, S.A. & Eva, A. (1988) Molecular cloning and characterization of the human dbl proto-oncogene: evidence that its overexpression is sufficient to transform NIH/3T3 cells. EMBO J. 7:2465–2473

Sadowski, I., Stone, J.C. & Pawson, T. (1986) A noncatalytic domain conserved among cytoplasmic protein-tyrosine kinases modifies the kinase function and transforming activity of Fujinami sarcoma virus $p130^{gag\text{-}fps}$. Mol. Cell. Biol. 6:4396–4408

Sigal, I.S. (1988) The *ras* oncogene, a structure and some function. Nature (London) 332:485

Sigal, I.S., Gibbs, J.B., D'Alonzo, J.S. & Scolnick, E.M. (1986) Identification of effector residues and a neutralizing epitope of Ha-*ras*-encoded p21. Proc. Natl. Acad. Sci. USA 83:4725–4729

Sigal, I.S., Marshall, M.S., Schaber, M.D., Vogel, U.S., Scolnick, E.M. & Gibbs, J.B. (1989) Structure-function studies of the *ras* protein. Cold Spring Harbor Symp. Quant. Biol. 53:863–869

Tanaka, K., Matsumoto, K. & Toh-e, A. (1989) *IRA1*: an inhibitory regulator of the RAS/cAMP pathway in *Saccharomyces cerevisiae*. Mol. Cell. Biol. 9:757–768

Trahey, M. & McCormick, F. (1987) A cytoplasmic protein stimulates normal N-*ras* p21 GTPase, but does not affect oncogenic mutants. Science 238:542–545

Trahey, M., Wong, G., Halenbeck, R., Rubinfeld, B., Martin, G.A., Ladner, M., Long, C.M., Crosier, W.J., Watt, K., Koths, K. & McCormick, F. (1988) Molecular cloning of two types of GAP complementary DNA from human placenta. Science 242:1697–1700

Vogel, U.S., Dixon, R.A.F., Schaber, M.D., Diehl, R.E., Marshall, M.S., Scolnick, E.M., Sigal, I.S. & Gibbs, J.B. (1988) Cloning of bovine GAP and its interaction with oncogenic ras p21. Nature (London) 335:90–93

Wang, H.R. & Parsons, J.T. (1989) Deletions and insertions within an amino-terminal domain of pp60$^{v\text{-}src}$ inactivate transformation and modulate membrane stability. J. Virol. 63:291–302

Molecular Mechanisms of G-Protein Activation

K.H. Jakobs[1], P. Gierschik[1], G. Hilf[1], C. Reithmann[1], D. Sidiropoulos[1], and T. Wieland[1]

1 Introduction

Signal-transducing guanine nucleotide-binding regulatory proteins (G-proteins) are heterotrimeric proteins, localized in or at the inner surface of plasma membranes. The function of these proteins is to couple receptors for extracellular signaling molecules such as neurotransmitters, hormones, and auto- and paracrine hormonal factors to intracellular signal-forming systems (for reviews, see Rodbell 1980; Stryer 1986; Stryer and Bourne 1986; Gilman 1987; Birnbaumer et al. 1987; Casey and Gilman 1988; Lochrie and Simon 1988; Neer and Clapham 1988; Chabre and Deterre 1989; Gierschik et al. 1989a). While the number of plasma membrane-located receptors interacting with and activating G-proteins is rather high, almost 100 such receptors are known at present, the number of G-proteins activated by these receptors, known so far from both the protein and cDNA level, is rather small, in the range of ten, with some additional G-proteins apparently occurring only in certain specialized cells. For example, the transducins (G_t), the G-proteins involved in the light signal transduction, are confined to specialized cells in the retina. On the other hand, other G-proteins such as the G_s- and G_i-proteins occur in almost any cell type, at least one of the subtypes of these proteins, four and three of which, respectively, are known so far. Other G-proteins such as the G_o-proteins are not confined to one specialized cell type but are mainly localized in neuronal cells.

The basic composition of the various signal-transducing G-proteins is apparently identical. These proteins are heterotrimeric proteins, with an α-subunit (M_r's on SDS PAGE of 39 to 52 kDa), which binds guanine nucleotides such as GDP and GTP and their respective analogs, a β-subunit, two of which (β_1, β_2) are known presently on the protein and cDNA level with M_r's of 36 and 35 kDa, respectively, and a γ-subunit with M_r's in the range of 8 to 10 kDa. The α-subunits are apparently the most distinct subunits in the various G-proteins, although large homologies are found in these subunits when comparing the different G-protein-subtypes. These homologies are even higher in a given G-protein subfamily, e.g., the G_s- or the G_i-protein family of G-proteins. In contrast, the β-subunits of G-proteins are apparently identical in the various G-proteins, although the relative distribution of the β_1- and β_2-subtypes in various cellular systems can be quite different. For example, in mammalian rod outer segments only the 36 kDa β-subunit is apparently present, while in other cell types both subunits are found together with the various G-protein α-subunits, with the relative

[1]Pharmakologisches Institut der Universität Heidelberg, Im Neuenheimer Feld 366, D-6900 Heidelberg, Fed. Rep. of Germany

40. Colloquium Mosbach 1989
Molecular Mechanisms of Hormone Action

distributions of the β_1- and β_2-subtypes being apparently more cell- and than G-protein-specific. The γ-subunits of G-proteins, the primary structure of only one of these subunits is reported so far (γ-subunit of transducin), are apparently also a class of proteins with relatively great heterogeneity. Under nondenaturating conditions, the γ-subunits are always found associated with the β-subunits as $\beta\gamma$-dimers. The apparent heterogeneity of $\beta\gamma$-complexes is so far not reflected in a great heterogeneity of functions of these complexes in G-protein action. Except for transducin, which is apparently only loosely associated with the rod outer segment membranes and which can be released from the membranes by reducing the ionic strength of the medium, the other G-proteins are integral membrane proteins. Furthermore, while transducin, at least its α-subunit, is released from the membranes in the course of its activation by the photoexcited receptor rhodopsin, the release of other G-protein α-subunits from membranes by agonist-activated receptors is still an open question.

2 The Activation – Deactivation Cycle of G-Proteins

The basic mechanisms of G-protein activation by agonist-liganded activated receptors are, at least for some G-proteins, e.g., G_s and G_t, relatively well established. Based on a large body of information, the following sequence of events is assumed to occur (see, e.g., Stryer 1986; Gilman 1987; Birnbaumer et al. 1987; for reviews): In the inactive state, GDP is bound to the G-protein α-subunit and, in addition, the G-protein $\beta\gamma$-subunits are associated with the α-subunits as holo-G-protein complexes. Agonist binding to receptors capable of interacting with a given G-protein induces a conformational change of the receptor, now binding to the holo-G-protein. This receptor-G-protein interaction induces a conformational change in the G-protein α-subunit, which results in a reduced affinity of the G-protein for bound GDP and subsequent dissociation of GDP from the G-protein. The dissociation of GDP, in turn, apparently also alters the interaction of the G-protein with the agonist-liganded receptor, leading to a further conformational change of the receptor. This alteration of the receptor is monitored as an increase in receptor affinity for the bound agonist. When GTP is present, it will now bind to the nucleotide-free G-protein α-subunit. The binding of GTP has apparently at least three consequences. First, the G-protein dissociates, at least functionally, from the receptor. Second, the G-protein dissociates into its subunits, the GTP-liganded α-subunit and the $\beta\gamma$-dimer. Third, the receptor, now dissociated from the G-protein, loses its affinity for the bound agonist, which, then, can dissociate from the receptor. The GTP-liganded α-subunit interacts with the effector moiety, responsible for intracellular signal formation, and changes its activity. The active state of the G-protein is terminated by hydrolysis of the bound GTP to GDP by a GTPase activity, being an intrinsic property of the G-protein α-subunit. Furthermore, the GDP-liganded α-subunit will reassociate with the $\beta\gamma$-dimer and, thus, the inactive holo-G-protein complex will be formed. This inactive complex can now reenter a new activation-deactivation cycle.

Although many of these assumed features of G-protein activation by agonist-liganded receptors appear to be well established, there is still a large number of problems unresolved. In this short review, only three specific questions will be addressed on which we have recently focused in our laboratory. First, what is the relationship between agonist-receptor binding and G-protein activation? Second, how many (different) G-proteins are activated by one agonist-activated receptor in the native membrane milieu? And third, is there a specific function of the $\beta\gamma$-dimer in G-protein activation by GTP in the presence of agonist-activated receptors?

3 Relationship Between Agonist-Receptor Binding and G-Protein Activation

The basic observations which lead to the hypothesis that G-protein activation is only by the high affinity agonist-binding receptor are the following (see, e.g., Lefkowitz et al. 1983; Gilman 1987; Gierschik et al. 1988; for reviews): First, in the physical or functional absence of G-proteins, only low affinity agonist receptor binding is observed. Second, as best studied with β- and α_2-adrenoceptors, there is a good correlation between the formation of the high affinity agonist-binding state of the receptor and the ability of various agonists (full and partial agonists) to activate G_s (β-adrenoceptors) and G_i (α_2-adrenoceptors) with subsequent stimulation and inhibition, respectively, of adenylate cyclase. The high affinity agonist-binding state of the receptor is, in most systems, dependent on the presence of divalent cations such as Mg^{2+} and Mn^{2+} at millimolar concentrations. Finally, activation of G-proteins by GTP and its analogs guanylyl-5′-imidodiphosphate (GppNHp) and guanosine-5′-0-(3-thiotriphosphate) (GTP[γS]) coincides with a reduction in receptor affinity for agonist, while antagonist receptor binding is usually not altered by the addition of guanine nucleotides. All of these data lead to the hypothesis that in the absence of guanine nucleotides the agonist-receptor-G-protein complex (ternary complex) always exhibits high affinity for the agonist and that only this high affinity agonist-binding complex is capable of activating G-proteins, i.e., inducing the dissociation of G-protein-bound GDP and the subsequent binding of GTP to the G-protein α-subunit.

In measuring binding of the formyl peptide receptor agonist, fMet-Leu-Phe (fMLP), to its binding sites in membranes of differentiated human leukemia cells (HL 60), we observed that the receptor exists in at least two affinity states for the agonist, with the appearance of the high affinity binding state being absolutely dependent on divalent cations at millimolar concentrations (Gierschik et al. 1989b). Most importantly, the addition of guanine nucleotides such as GTP, GTP[γS], GDP or guanosine-5′-0-(2-thiodiphosphate) (GDP[βS]) not only reduced the high affinity agonist receptor binding (by more than 90%), these nucleotides also decreased agonist receptor binding in the complete absence of divalent cations, thus, at a receptor exhibiting low agonist affinity. These data suggest that the interaction of the agonist-liganded receptor with the G-protein does not necessarily induce a high affinity binding state of the receptor for the agonist. As agonist receptor binding on adrenoceptors (the model system) is generally studied as agonist competition of labeled antagonist binding (Lefkowitz et al. 1983), the regulation of low affinity agonist binding by guanine nucleotides is, due to technical reasons, not detectable. Although the relative extent of regulation of low affinity agonist receptor binding by guanine nucleotides is rather high, e.g., guanine nucleotides can decrease fMLP receptor binding by more than 60% under this condition, the absolute extent is rather small, about one-tenth, compared to the guanine nucleotide regulation of high affinity receptor binding.

In the low agonist affinity state, the receptor not only interacts with a G-protein but is also capable of activating the G-protein. Using submicromolar concentrations of free Mg^{2+}, which are necessary but also sufficient for providing agonist-stimulated GTPase activity of G-proteins, which however do not induce high affinity agonist receptor binding, we demonstrated that under this condition the agonist-activated formyl peptide receptor can perfectly stimulate a high affinity GTPase in HL 60 membranes (Gierschik et al. 1989b). By increasing the free Mg^{2+} to millimolar concentrations and, thereby, inducing the high affinity agonist-binding state of the receptor, we only observed an increase in the potency of the agonist to stimulate the G-protein-dependent GTP hydrolysis. There was an excellent correlation between the actual high and low affinity dissociation constants for agonist receptor binding and the EC_{50} values for the agonist to stimulate GTPase activity. Furthermore, inclusion of any agent, which reversed the high affinity agonist-binding state of the receptor, such as sodium ions (Gierschik et al. 1989c), GDP, or high GTP concentrations, in the GTPase assay decreased the potency of the agonist to stimulate GTPase activity. Finally, when a GTPase assay was performed in the presence of millimolar concentrations of Mg^{2+} and with a very low GTP concentration, which had no effect on agonist receptor binding, GTPase stimulation was apparently by both the high and low affinity agonist-binding receptor (Gierschik and Jakobs in preparation).

Thus, in summary to this point, interaction of agonist-liganded receptors with G-proteins does not necessarily induce high affinity agonist receptor binding. The low affinity state of the receptor is apparently capable of both interacting with and activating G-proteins. It has to be pointed out here that this low affinity agonist binding state of the receptor is not identical to the state observed in the absence of G-proteins or when the interaction of receptors with G-proteins is prevented. Thus, there are apparently at least three states of the receptor with regard to its binding affinity for the agonist, a very low affinity state (physical or functional absence of G-proteins), a low affinity state (usually observed in the absence of divalent cations), and a high affinity agonist-binding state (usually requiring millimolar concentrations of divalent cations). In the latter two states, the agonist-liganded receptor can interact with and activate G-proteins. Whether in intact cells, G-protein activation is induced by the high and/or low affinity agonist-binding state of the receptor is more difficult to answer, considering on the one hand the rather high ambient Mg^{2+} concentration and, on the other hand, the rather high level of the endogenous guanine nucleotides GTP and GDP.

4 Amplification and Diversification of Signal Transduction at the Receptor-G-Protein Level

From the vision system, it is well known that one photoexcited rhodopsin can activate many transducins, which is considered as one of the essential amplification steps in the light signal transducation cascade (Stryer 1986; Stryer and Bourne 1986; Chabre and Deterre 1989). Furthermore, when analyzing G_s-protein activation by β-adrenoceptors in systems reconstituted from purified components in phospholipid vesicles, it has been observed that one agonist-activated receptor is capable of activating several G_s-proteins (Pedersen and Ross 1982). Since the vision system is a specialized system both with regard to the stoichiometry of receptors and G-proteins and the actual topography of these two components and since the relative concentrations of the two components in reconstituted systems are rather arbitrarily chosen, we studied whether in an intact membrane milieu, which probably exhibits different constraints than those from the reconstituted system or the vision system, one agonist-activated receptor can also activate many G-proteins. For this, we used the above-mentioned HL 60 membrane system and measured the agonist (fMLP)-stimulated binding of the hydrolysis-resistant GTP analog GTP[γS] to G-proteins. Compared to the formyl peptide receptor number present in these membranes (1–2 pmol mg^{-1} protein), the agonist-activated receptor was capable of inducing the binding of GTP[γS] to about 20–30 pmol G-proteins mg^{-1} membrane protein (Moghtader et al. 1988; Gierschik et al. in preparation). Thus, even in an intact membrane milieu, one agonist-activated receptor is capable of activating several G-proteins. In real time analysis of agonist receptor binding and G-protein activation, it has to be determined whether, in fact, one receptor activated by one agonist molecule activates many G-proteins. In different cell types and with different receptors, apparently different degrees of amplification can be obtained. For example, in porcine atrial membranes, we observed "only" a two- to threefold amplification when we studied G-protein activation by muscarinic acetylcholine receptors (Hilf et al. 1989).

At the level of receptor-G-protein interaction, not only an amplification of signal transduction takes place but apparently also a diversification. There is functional evidence, for example, that the agonist-activated β-adrenoceptor can interact with both G_s and G_i-proteins (Asano et al. 1984; Jakobs et al. 1985). In addition, activation of two unidentified G-proteins by a single population of angiotensin II receptors has been suggested for hepatocyte membranes (Bouscarel et al. 1988). In HL 60 membranes, at least two distinct G_i-protein subtypes are present, G_{i2} and G_{i3}, with G_{i2} being the major form. We have recently reported that activation of formyl peptide receptors by its agonist fMLP substantially increases the ADP-ribosylation of G_i-proteins by cholera toxin in HL 60 membranes (Gierschik and Jakobs 1987). These data suggested, as also reported for ADP-ribosylation of transducin by cholera toxin (see Stryer 1986; for review), that by the interaction of the G-protein with the activated receptor the G-protein's conformation is altered, now being a substrate for cholera toxin. Using this methodology, which, thus, provides direct evidence for coupling of a given G-protein subtype to a given receptor, we have recently observed that the agonist-liganded formyl peptide receptor interacts with both G_{i2} and G_{i3} (Gierschik et al. 1989d). There was no difference in the concentrations of the peptide required to half-maximally and maximally stimulate ADP-ribosylation of the α-subunits of G_{i2} and G_{i3} by cholera toxin. These findings suggest that one rather than two distinct formyl peptide receptors interact with the two distinct G-proteins. It is likely that these two G-proteins when activated by formyl peptide receptors, in turn, regulate distinct intracellular signaling systems.

Thus, in summary to the second point, at the level of receptor-G-protein interaction, not only an amplification of the signal transduction cascade takes place but apparently also a diversification, in that one agonist-activated receptor is capable of not only activating many identical G-proteins but apparently also distinct G-proteins. This amplification and diversification, which are probably different for different cells and receptors, may explain the nonuniform response of various cell types to a given receptor or, on the other hand, the nonuniform response of one cell type to various receptors assumed to act via the same signal-transduction system.

5 Involvement of $\beta\gamma$-Subunits in G-Protein Activation

Activation of G-proteins by agonist-activated receptors requires both the guanine nucleotide-binding and GTP-hydrolyzing α-subunits and the $\beta\gamma$-dimers (for review, see, e.g., Gilman 1987). However, the role of the $\beta\gamma$-subunits in G-protein activation by agonist-liganded receptors is presently very hypothetical. It is assumed that the more hydrophobic $\beta\gamma$-subunits are required to keep the more hydrophilic α-subunits at the membrane and, thus, in touch with the receptor. Although such a role of the $\beta\gamma$-subunits is feasible, we were interested whether the $\beta\gamma$-dimers may have an additional, more specific function in the activation process of G-protein α-subunits by GTP and agonist-liganded receptors. Support for such a specific role of $\beta\gamma$-subunits in G-protein activation was obtained when we studied the effects of $\beta\gamma$-dimers of transducin purified by two distinct methods. When the $\beta\gamma$-subunits were prepared from transducin eluted from bovine rod outer segment membranes in the presence of GTP, the $\beta\gamma$-subunits added to HL 60 membranes had no effect, over a wide range of concentrations, on G-protein activity. In contrast, $\beta\gamma$-subunits purified from transducin eluted with the GTP analog GTP[γS] caused a marked G_i-protein activation when added to HL 60 membranes. This $\beta\gamma$-subunit activity was heat-sensitive and not due to a contamination by free GTP[γS] (Wieland and Jakobs 1989). We were able to convert an "inactive" $\beta\gamma$-subunit purified from transducin eluted with GTP into an "active" $\beta\gamma$-subunit by recombining the $\beta\gamma$-subunits with illuminated transducin-depleted rod outer segment membranes and GTP[γS]. Whereas the respective controls (minus rod outer segment membranes, minus GTP[γS] and rod outer segment membranes plus GTP[γS] alone) had virtually no effect, the $\beta\gamma$-subunits treated with photoexcited rhodopsin and GTP[γS] potently and efficiently activated G_i-proteins in HL 60 membranes. Furthermore, the thus "activated" $\beta\gamma$-subunits, but not the respective controls, caused an increase in G_s-protein activity when added to human platelet membranes with subsequent increase in adenylate cyclase activity (Wieland and Jakobs in preparation). When using [^{35}S]GTP[γS] instead of unlabeled GTP[γS], we observed that the β-subunit of transducin was thiophosphorylated. Finally, addition of an apparently thiophosphorylated β-subunit of transducin to HL 60 membranes in the presence of [^{3}H]GDP leads to the formation of [^{3}H]GTP[γS] (Wieland and Jakobs 1989).

Similar β-subunit (thio)phosphorylations as observed with the β-subunit of transducin were found by studying phosphorylations of HL 60 membrane proteins by [γ-^{32}P]GTP and [^{35}S]GTP[γS] (Wieland et al. in preparation). This phosphorylation was specifically inhibited by unlabeled GTP and GTP[γS] but only very weakly or not all by ATP, GppNHp, adenosine-5′-O-(3-thiotriphosphate), and adenylyl-5′-imidodiphosphate. Furthermore, when the membranes were first phosphorylated by GTP, addition of GDP but not of GDP[βS] led to the disappearance of the phosphorylated band. Finally, coaddition of GDP and the formyl peptide receptor agonist fMLP led to the increased formation of $[^{32}P]_i$; these findings suggest that the apparently energy-rich phosphate group bound to the β-subunit was transferred to GDP leading to the formation of [γ-^{32}P]GTP and that this labeled GTP was subsequently bound to the G-protein α-subunit and finally hydrolyzed by the GTPase activity of the G-protein to GDP and $[^{32}P]_i$.

Thus, in summary, although many details of the above described reactions have not yet been elucidated, we would like to propose that the βγ-subunits of G-proteins are not merely membrane anchors for the α-subunits but that specifically the β-subunits have a distinct function in G-protein activation by agonist-liganded receptors and GTP or GTP[γS], by acting as intermediate (thio)phosphate group transfer proteins. Whether such phosphate transfer reactions, for which indirect functional evidence has recently been provided (Jakobs and Wieland 1989), are obligatory for G-protein activation by GTP and agonist-liganded receptors or alternative reactions remains to be determined.

6 Conclusions

G-proteins serve as intermediates in signal transduction in a wide variety of systems by coupling receptors for a highly diverse group of extracellular signaling molecules to various intracellular signal-forming systems. The elucidation of the structural features of these coupling components revealed a much greater diversity of the individual G-proteins and their subunits than previously imagined. Furthermore, the mechanisms by which these proteins are activated by agonist-liganded receptors and, which thus transduce and amplify the signals, are still far from being clarified, particularly the role of the individual G-protein subunits. Our actual ignorance of these proteins and their functions at the cellular level may even be much higher considering not only the increasing number of G-proteins and G-protein-coupled receptors but also the rapidly increasing number of effector systems apparently regulated in their activity by G-proteins.

Acknowledgments. The authors' studies reported herein were supported by the Deutsche Forschungsgemeinschaft, the Fritz Thyssen-Stiftung, and the Fonds der Chemischen Industrie.

References

Asano, T., Katada, T., Gilman, A.G. & Ross, E.M. (1984) Activation of the inhibitory GTP-binding protein of adenylate cyclase, G_i, by β-adrenergic receptors in reconstituted phospholipid vesicles. J. Biol. Chem. 259:9351–9354.

Birnbaumer, L., Codina, J., Mattera, R., Yatani, A., Scherer, N., Toro, M.-J. & Brown, A.M. (1987) Signal transduction by G proteins. Kidney Int. 32 (Suppl. 23):S-14–S-37.

Bouscarel, B., Blackmore, P.F. & Exton, J.H. (1988) Characterization of the angiotensin II receptor in primary cultures of rat hepatocytes. Evidence that a single population is coupled to two different responses. J. Biol. Chem. 263:14913–14919.

Casey, P.J. & Gilman, A.G. (1988) G Protein involvement in receptor-effector coupling. J. Biol. Chem. 263:2577–2580.

Chabre, M. & Deterre, P. (1989) Molecular mechanism of visual transduction. Eur. J. Biochem. 179:255–266.

Gierschik, P. & Jakobs, K.H. (1987) Receptor-mediated ADP-ribosylation of a phospholipase C-stimulating G protein. FEBS Lett. 224:219–223.

Gierschik, P., McLeish, K. & Jakobs, K.H. (1988) Regulation of G-protein-mediated signal transfer by ions. J. Cardiovasc. Pharmacol. 12 (Suppl. 5):S20–S24.

Gierschik, P., Sidiropoulos, D., Dieterich, K. & Jakobs, K.H. (1989a) Structure and function of signal-transducing, heterotrimeric GTP-binding proteins. In: Habenicht, A. (ed) Growth factors, differentiation factors and cytokines. Springer, Berlin Heidelberg New York Tokyo (in press)

Gierschik, P., Steißlinger, M., Sidiropoulos, D., Herrmann, E. & Jakobs, K.H. (1989b) Dual Mg^{2+} control of formyl peptide receptor-G-protein interaction in HL 60 cells: evidence that the low agonist affinity receptor interacts with and activates the G-protein. Eur. J. Biochem. 183:97–105.

Gierschik, P., Sidiropoulos, D., Steißlinger, M. & Jakobs, K.H. (1989c) Na^{+} regulation of formyl peptide receptor-mediated signal transduction in HL 60 cells: evidence that the cation prevents unoccupied receptors from activating the G-protein. Eur. J. Pharmacol. Mol. Pharmacol. (in press)

Gierschik, P., Sidiropoulos, D. & Jakobs, K.H. (1989d) A receptor not coupled to adenylate cyclase interacts with two distinct G_i-proteins in the same cell. (submitted)

Gilman, A.G. (1987) G proteins: transducers of receptor-generated signals. Annu. Rev. Biochem. 56:615–649.

Hilf, G., Gierschik, P. & Jakobs, K.H. (1989) Muscarinic acetylcholine receptor-stimulated binding of guanosine-5′-O-(3-thiotriphosphate) to G-proteins in cardiac membranes. (submitted)

Jakobs, K.H. & Wieland, T. (1989) Evidence for receptor-regulated phosphotransfer reactions involved in activation of the adenylate cyclase inhibitory G-protein in human platelet membranes. Eur. J. Biochem. 183:115–121.

Jakobs, K.H., Aktories, K., Minuth, M. & Schultz, G. (1985) Inhibition of adenylate cyclase. Adv. Cyclic Nucl. Protein Phosphoryl. Res. 19:137–150.

Lefkowitz, R.J., Stadel, J.M. & Caron, M.G. (1983) Adenylate cyclase-coupled beta-adrenergic receptors: structure and mechanisms of activation and desensitization. Annu. Rev. Biochem. 52:159–186.

Lochrie, M.A. & Simon, M.I. (1988) G Protein multiplicity in eukaryotic signal transduction systems. Biochemistry 27:4957–4965.

Moghtader, R., Beck, M., Büch, M. & Straub, C. (1988) Regulation of guanosine-5′-O-(3-thiotriphosphate) binding to G proteins in membranes of HL 60 cells. Naunyn-Schmiedeberg's Arch. Pharmacol. 337 (Suppl.):R37.

Neer, E.J. & Clapham, D.E. (1988) Roles of G protein subunits in transmembrane signalling. Nature (London) 333:129–134.

Pedersen, S.E. & Ross, E.M. (1982) Functional reconstitution of β-adrenergic receptors and the stimulatory GTP-binding protein of adenylate cyclase. Proc. Natl. Acad. Sci. USA 79:7228–7232.

Rodbell, M. (1980) The role of hormone receptors and GTP-regulatory proteins in membrane transduction. Nature (London) 284:17–22.

Stryer, L. (1986) Cyclic GMP cascade of vision. Annu. Rev. Neurosci. 9:87–119.

Stryer, L. & Bourne, H.R. (1986) G proteins: a family of signal transducers. Annu. Rev. Cell Biol. 2:391–419.

Wieland, T. & Jakobs, K.H. (1989) Mechanisms of G-protein activation by GTP[γS]: the β-subunit of transducin serves as a thiophosphorylated intermediate leading to G-protein activation in HL 60 membranes. Naunyn-Schmiedeberg's Arch. Pharmacol. 339 (Suppl.):R33.

Interaction of Transducin with Retinal cGMP Phosphodiesterase

PH. DETERRE[1] and M. CHABRE[1]

1 Transducin and PDE: Two Multisubunit Peripheral Proteins

Transducin is a member of the large family of heterotrimeric G-proteins that convey hormonal or sensory signals from transmembranous receptors to various types of effector proteins. It is a peripheral membrane protein, bound to the cytoplasmic face of the membrane from which it can be detached easily by lowering the ionic strength of the surrounding medium (see Kühn 1984 for a review). The three polypeptides form only two functional subunits, T_{α} (39 kDa) and $T_{\beta\gamma}$ (36 + 6 kDa), as T_{β} and T_{γ} cannot be dissociated without denaturation. T_{α} bears a guanine nucleotide binding site that is highly conserved among all G-proteins. In the resting state, i.e., in the absence of photoexcited rhodopsin, a GDP, practically inexchangeable on a time scale of hours, remains locked in this nucleotide site (Bennett and Dupont 1985). T_{α}GDP, associated to $T_{\beta\gamma}$, diffuses freely on the membrane surface. In the "activated state", i.e., with a bound GTP, T_{α}GTP is dissociated from $T_{\beta\gamma}$. The β-polypeptide of transducin is strictly identical to the 36 kDa β-polypeptide of Gs or Gi. Differences between the γ-polypeptides must therefore account for the fact that isolated $T_{\beta\gamma}$ is only weakly membrane attached and partially soluble even in a physiological medium, while the other $G_{\beta\gamma}$ appear very strongly membrane attached.

A major class of the G-protein effectors is that acting on the intracellular pool of cyclic nucleotides, the most-studied example being the adenylate cyclase that can be activated or inhibited, respectively, by Gs or Gi. In the visual transduction cascade the soluble internal messenger is the cyclic guanosine monophosphate (cGMP) and the modulation by light is negative through an activation of an enzyme degrading the messenger, the retinal cGMP phosphodiesterase (PDE) (see Chabre and Deterre 1989 for a recent review on the visual transduction cascade). In the first step of the cascade, an inactive holotransducin T_{α}GDP-$T_{\beta\gamma}$ binds to the photoexcited rhodopsin (R*). The interaction with R* "opens" the nucleotide site in Tα, allows the fast release of the previously locked-in GDP so that a GTP can enter the empty site. As for all G-proteins, the binding of GTP triggers in the T_{α} subunit a conformational change that induces its release from the receptor, rhodopsin, and also its dissociation from the other transducin subunit $T_{\beta\gamma}$. In the retinal system the dissociation is made evident by the fact that T_{α}GTP becomes soluble even in a physiological medium and is released from the membrane on which $T_{\beta\gamma}$ remains mostly bound. T_{α}GTP alone seems responsible for the activation of the PDE, with which $T_{\beta\gamma}$ has apparently no direct interaction.

[1]Laboratoire de Biophysique Moléculaire et Cellulaire, Unité Associée 520 du CNRS, Fédération de Biologie, DRF/CENG, BP 85, F38041 Grenoble, France

40. Colloquium Mosbach 1989
Molecular Mechanisms of Hormone Action

By contrast with the Gs activable adenylate cyclase, which has an intrinsic transmembrane attachment, the retinal PDE is a peripheral membrane-bound enzyme (Baehr et al. 1979). The catalytic complex is made up of two undissociable, nearly identical subunits PDE_α (88 kDa) and PDE_β (84 kDa). The basal activity is very low. It was observed very early that a very high PDE activity could be revealed by proteolysis (Miki et al. 1975): in the resting state the activity of the $PDE_{\alpha\beta}$ complex is blocked by inhibitory subunit(s), that trypsin can degrade much faster than it can attack the active site of the catalytic subunits. On the other hand, the inhibitor resists a high temperature acidic treatment that denaturates irreversibly the catalytic complex. This technique was used to isolate and purify an 11 kDa subunit (denoted I) that was shown to reinhibit trypsin-activated $PDE_{\alpha\beta}$ (Hurley and Stryer 1982).

2 How Does Transducin Activate the PDE?

The fact that the PDE can be activated by proteolysis of the inhibitor suggested that the natural activation mechanism of the PDE by transducin results from an interaction between T_αGTP and the PDE inhibitor, which would force the release of the inhibitory constraint of I on $PDE_{\alpha\beta}$. The mechanism through which the inhibition is relieved remained, however, uncertain: it was not clear whether T_α associated to the PDE complex (as apparently Gs_α does with the adenylate cyclase) and released the inhibition without physically removing the inhibitor, or whether T_α interacted only with the inhibitor and formed a complex that would dissociate from $PDE_{\alpha\beta}$. Would this complex then remain membrane-bound or follow T_α in the solution? The number of inhibitory subunits per catalytic complex was also unknown, it was only a simple assumption that a single inhibitory subunit blocked the activity of the dimeric catalytic complex and that the activation would be an all-or-none process correlated with the removal of this unique inhibitor (Wensel and Stryer 1986).

Our approaches to these problems (Deterre et al. 1986, 1988) were based mainly on biochemical separation by anion exchange chromatography, and characterization of the various protein complexes that could be eluted, under various ionic conditions, from retinal rod membranes on which "activated" transducin and PDE had been allowed to interact. But in the physiological cycle the spontaneous hydrolysis of GTP in T_α soon regenerates T_αGDP conformation that loses its activating interaction on PDE and recovers a high affinity to rebind to $T_{\beta\gamma}$. There are good arguments, but no definite proof to date, that the rate of hydrolysis is independent of the eventual interaction of T_αGTP with PDE. In any case, this interaction is not expected to stabilize the T_αGTP form, whose half-life is only a few seconds for free transducin (Sitarramaya et al. 1988; Deterre, Vuong, Pagès and Chabre, unpublished results). The study of the interaction of T_α with PDE is therefore easier when the "active" conformation of T_α is made quasi-permanent by the use of a nonhydrolyzable analog of GTP, like GTPγS, that binds with a comparable affinity and induces in situ a permanent activation of PDE.

3 Identification of the T_αGTPγS-I Complex

A first set of experiments demonstrated that upon illumination of rod cell homogenates in the presence of GTPγS and in a physiological medium, a fraction of the T_αGTPγS produced is retained on the membrane. This retention is correlated to the presence of the PDE. However, in a low ionic strength medium, all of the transducin and the PDE are extracted. In the elution profile of this extract on an ion exchange column, a small peak containing both T_α and the PDE inhibitor polypeptide appeared before the main peaks of T_α, T_β, and PDE, (Fig. 1). Gel filtration studies confirmed that a low molecular weight complex containing T_α and I, and devoid of PDE catalytic subunits, existed already in the extract. The relative titration of the two components in the isolated complex, and its elution volume on the gel filtration column, strongly suggested that it results from a monomeric and 1:1 stoichiometric association.

4 Membrane Attachment of the T_αGTPγS-I Complex

The complex that we characterized has been artificially extracted from the membrane by low ionic strength elution. If the ionic composition of the medium is kept close to the physiological one (100 mM KCl + NaCl, 1 mM $MgCl_2$), no inhibitor is found with the soluble T_αGTPγS in the supernatant of pelleted membranes. Instead, a significant fraction of the T_αGTPγS pool, which depends linearly on the amount of native PDE initially present on the membrane, remains membrane-bound. Light-scattering studies, which allow the monitoring of the instant release of T_αGTP in the solution (Kühn et al. 1981), confirmed that addition of excess native PDE to the membrane suspension before illumination induces a proportional decrease in the amount of T_αGTP released in solution in a physiological medium. However, the fraction of T_αGTPγS retained on the membrane at physiological ionic strength is larger than that found associated to PDE inhibitor in the elution profile of the corresponding low ionic strength extract. This indicates that the low ionic strength manipulation required to extract the complex from the membrane also dissociates T_αGTPγS partially from I. Hence, the yield of the T_α-I complex extracted at low ionic strength cannot be used to estimate the amount of binding of T_αGTPγS to the inhibitor in situ. The method also does not allow to check whether the T_α-I complexes on the membrane are physically dissociated from the catalytic complexes of the PDE. The fact, however, that these two components are dissociated when in solution, in a low ionic strength medium, suggests that they are not tightly interacting when bound to the membrane in a physiological medium.

5 How Many Inhibitors Regulate the Activity of the PDE?

Under the usual assumption of a 1:1:1 relative stoichiometry of the α,β, and I subunits in the native PDE, one would expect the activated PDE resulting from the removal of inhibitor by T_α to be devoid of I. We wondered whether this activated PDE could be separated from the native one. Indeed, on the ion exchange elution profile of the GTPγS-activated total extract, in addition to the peak observed corresponding to that of inactivated PDE, a second peak containing the PDE catalytic units was detected, which eluted at a higher salt concentration. But to our surprise this peak contained, besides PDE_α and PDE_β, a significant amount of inhibitory subunit. Further modifications of the gradient elution program then allowed the resolution of this second peak into two components (Figs. 1 and 2): a major one in which the ratio of $I/PDE_{\alpha\beta}$ was about one-half that measured in the native PDE peak, and a minor one that appeared totally devoid of I. Both components corresponded to active PDE, but the specific activity (i.e., the amount of catalytic subunits) is approximately doubled for

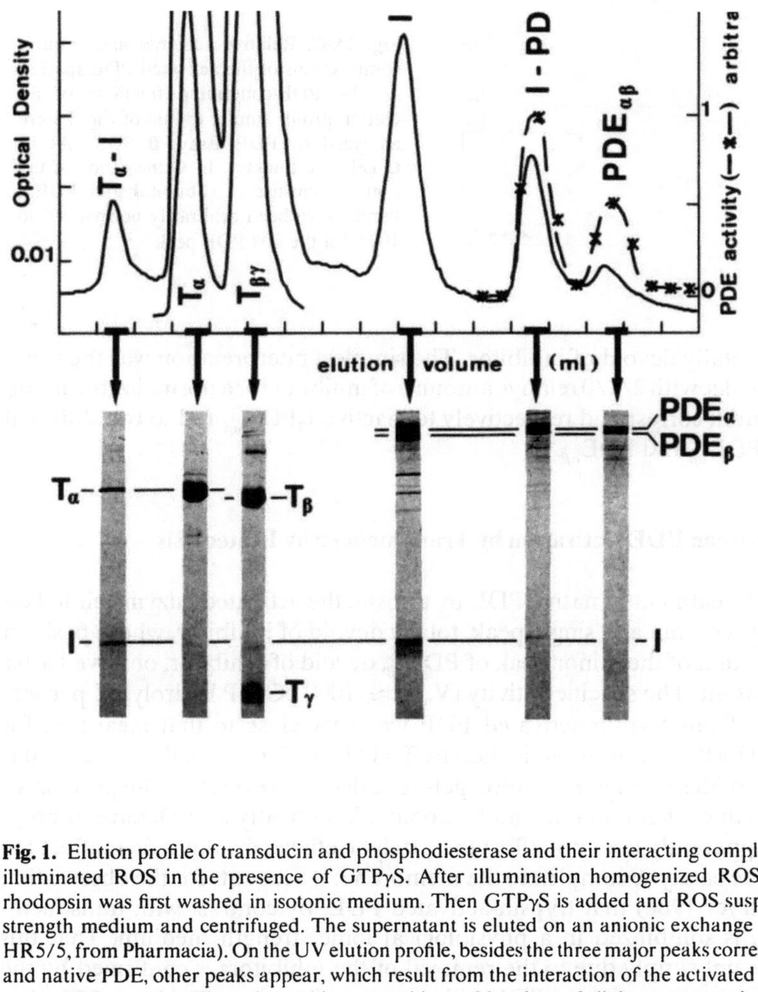

Fig. 1. Elution profile of transducin and phosphodiesterase and their interacting complexes extracted from illuminated ROS in the presence of GTPγS. After illumination homogenized ROS containing 15 mg rhodopsin was first washed in isotonic medium. Then GTPγS is added and ROS suspended in low ionic strength medium and centrifuged. The supernatant is eluted on an anionic exchange column (Polyanion HR5/5, from Pharmacia). On the UV elution profile, besides the three major peaks corresponding to T_α, $T_{\beta\gamma}$, and native PDE, other peaks appear, which result from the interaction of the activated transducin with the native PDE complex. The polypeptide compositions of fractions of all these separated peaks were analyzed by SDS PAGE (*below*)

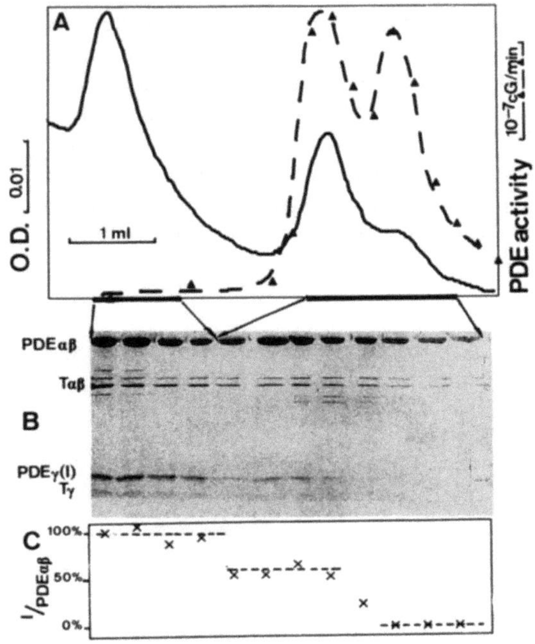

Fig. 2A-C. Relative activities and subunit compositions of the activated PDE species. **A** The PDE-containing fractions of an elution profile similar to that of Fig. 1 were analyzed by PDE assay; **B** SDS PAGE; **C** gel densitometry. In **C** the ratios of the optical densities of subunit I and $PDE_{\alpha\beta}$ bands have been arbitrarily normalized to 100% for the first PDE peak

the component totally devoid of inhibitor. The simplest interpretation was then that the three PDE peaks, with 2/1/0 relative amounts of inhibitor for a normalized amount of catalytic subunit, correspond respectively to inactive $I_2PDE_{\alpha\beta}$ and to two different active states, I-$PDE_{\alpha\beta}$ and $PDE_{\alpha\beta}$.

6 Analogies Between PDE Activation by Transducin or by Proteolysis

Under the usual treatment of native PDE by trypsin, the activated enzyme eluted on the ion exchange column as a single peak, totally devoid of inhibitor, whose position corresponded to that of the minor peak of $PDE_{\alpha\beta}$, devoid of inhibitor, observed after $T_\alpha GTP\gamma S$ activation. The specific activity ($V_{max} \approx$ 1000 cGMP hydrolyzed per second per PDE) of this trypsin-activated PDE was very close to that measured for $PDE_{\alpha\beta}$ obtained by the "natural" activation by $T_\alpha GTP\gamma S$. Careful analysis of the catalytic unit polypeptides on high resolution gels revealed, however, that the proteolysis had trimmed both of them on one end by about 2 kDa (Catty and Deterre in preparation). This apparently does not affect the catalytic efficiency, but it does affect the attachment of the catalytic complex to the membrane: it accounts for the observation (Wensel and Stryer 1986) that trypsin-activated PDE, by contrast with transducin-activated PDE, is solubilized in a physiological ionic strength medium. That this solubilization is not simply due to the removal of the inhibitors is confirmed by the observation that trypsin-activated PDE will not reattach to the membrane even after its total inhibition by addition of an excess of purified inhibitor.

The progressive effect of mild tryptic digestion at 0°C was then scrutinized: after a very short digestion time the native PDE peak decreases, first to give rise to a major peak that corresponds to the I-$PDE_{\alpha\beta}$ observed with T_αGTPγS, and a minor peak of PDE totally devoid of inhibitor. Then, only upon longer digestion does the native PDE disappear completely and a third peak appears that corresponds to totally stripped $PDE_{\alpha\beta}$. Still later, the I-$PDE_{\alpha\beta}$ peak decreases in turn and is quantitatively converted into the most active one. The specific activities in solution of the two trypsinized PDE species are identical to that found for the corresponding transducin-activated species. But, by contrast with transducin-induced activation, which under our conditions always produced only a small proportion of totally stripped $PDE_{\alpha\beta}$, tryptic activation leads easily to practically pure $PDE_{\alpha\beta}$. This accounts for the common observation that the specific activity of PDE obtained in vitro by the action of T_αGTPγS is much lower than that obtained by proteolysis: the concentrations of transducin commonly used in vitro are much lower than that reached in vivo in a retinal rod, and must be too low to strip a significant proportion of the PDE of its second inhibitor. As T_αGTPγS and T_αGTP are soluble, their action depends, contrary to what is often wrongly assumed, on their concentration in the solution rather than on their stoichiometric ratio with respect to native PDE.

The observation that the active states of the PDE generated by progressive proteolysis are kinetically identical to that obtained by interaction with transducin strongly supports the concept that transducin activates the PDE by physically removing the inhibitory subunits. The occurrence of two steps to reach the fully active PDE indicates that the two inhibitory subunits are taken off one at a time, which confirms that one T_αGTPγS binds stoichiometrically to one inhibitory subunit.

7 Possible Roles of the Two Inhibitors in the Regulation of PDE Activity by Transducin

Our method of separation of the different PDE complexes by elution on an ion exchange column does not give direct access to the relative affinities of the inhibitors for the catalytic complexes and transducin, as dissociation or recombination may occur on the column during the loading and/or the elution process. We observed, however, that the concentration of transducin in the mixture before the elution influences the relative yields of the two activated PDE species.

Evidence for rapid exchange of inhibitor subunits between native I_2-$PDE_{\alpha\beta}$ and fully activated $PDE_{\alpha\beta}$ was obtained by mixing aliquots of these two complexes after their separation on the column, and eluting the mixture again once more on the same column: a large peak of I-$PDE_{\alpha\beta}$, produced by the exchange reaction $I_2PDE_{\alpha\beta} + PDE_\beta \leftrightarrows 2$ I-$PDE_{\alpha\beta}$, at the expense of both initial components, was observed in the elution profile. The occurrence of this exchange obscures the estimate of the activity of the intermediate species I-$PDE_{\alpha\beta}$, since the purified species can partially transform in the two other ones. The ratio of the V_{max} for $PDE_{\alpha\beta}$ and I-$PDE_{\alpha\beta}$ was always found between 2 and 2.5, both species having a high specific activity in the range of 10^3 cGMP hydrolyzed per second per PDE at high cGMP concentration, i.e., in the millimolar cGMP range. The Km of $PDE_{\alpha\beta}$ evidently lies at a much lower concentration, close to the physiologically relevant range around 1 to 10 μM. But the Km of I-$PDE_{\alpha\beta}$ is very hard to measure: one cannot dilute infinitely this species without risking the dissociation of the inhibitor and creating some $PDE_{\alpha\beta}$. Then, due to the high specific activity, the cGMP consumption remains too high to allow a rate measurement at a stable cGMP concentration in the micromolar range. The issue is, however, capital: a very interesting regulation model could be based on the speculation that the specific activity of the intermediate species I-$PDE_{\alpha\beta}$ is significantly reduced when the cGMP concentration reaches a physiological range. The model is based on the inhibitor exchangeability demonstrated above and could account for the rapid termination of the physiological response to a weak light flash in a rod. This fast termination, within a few 100 ms, is not due to an inactivation of the cGMP-sensitive cationic channels. It, therefore, requires that every step in the cGMP cascade be rapidly blocked after the flash. Photoexcited rhodopsin itself is rapidly inhibited by phosphorylation followed by arrestin binding (Wilden et al. 1986). However, for decoupling transducin from the PDE inhibitor, the GTPase rate seems too slow to account for the turn off that would be correlated with the decay of T_αGTP in T_αGDP, which would release the PDE inhibitor. We have already pointed out that in vivo the T_αGTP concentration becomes locally very high around the photoexcited rhodopsin, and that this high concentration seems to be required to produce the most active $PDE_{\alpha\beta}$. But the T_αGTP, which are very soluble, rapidly dilute and after blocking the photoexcited rhodopsin their concentration may rapidly become too low to further produce fully stripped $PDE_{\alpha\beta}$. The initially formed $PDE_{\alpha\beta}$ may diffuse also on the membrane into areas where they encounter excess native I_2-$PDE_{\alpha\beta}$. Exchange of inhibitor will then convert all the $PDE_{\alpha\beta}$ into I-$PDE_{\alpha\beta}$. If the Km of I-$PDE_{\alpha\beta}$ for cGMP were significantly higher than that of $PDE_{\alpha\beta}$ and above the physiological range of regulation of the cationic channels, this diffusion and exchange process could allow a rapid quenching of the PDE activity at the appropriate cGMP concentration. This model is still very speculative and attempts at measurement of the specific activity of I-$PDE_{\alpha\beta}$ at low cGMP concentration are in progress. Whatever the result, it is likely that the occurrence of multiple inhibitor subunits is related to subtle regulation of the PDE activity by transducin.

In the cyclase system for comparison, evidence shows that a G_αGTP subunit remains attached to the catalytic complex that it activates: there is no separable inhibitory subunit, but more likely an inhibitory domain that might be displaced but not removed by its interaction with Gs_αGTP. There is, as yet, no evidence suggesting multisubunit attachments, which correlates with the fact that the cyclase seems to possess a single catalytic unit.

References

Baehr, W., Devlin, M.J. & Applebury, M.L. (1979) Isolation and characterization of cGMP phosphodiesterase from bovine rod outer segments. J. Biol. Chem. 254:11669–11677

Bennett, N. & Dupont, Y. (1985) The G-protein of retinal rod outer segments (transducin). J. Biol. Chem. 260:4156–4168

Chabre, M. & Deterre, P. (1989) Molecular mechanism of visual transduction. Eur. J. Biochem. 179:255–266

Deterre, P., Bigay, J., Robert, M., Pfister, C., Kühn, H. & Chabre, M. (1986) Characterization of the complex formed by phosphodiesterase inhibitor and transducin α subunit. Proteins Struct. Funct. Genet. 1:188–193

Deterre, P., Bigay, J., Forquet, F., Robert, M. & Chabre, M. (1988) cGMP phosphodiesterase of retinal rods is regulated by two inhibitory subunits. Proc. Natl. Acad. Sci. USA 85:2424–2428

Hurley, J. & Stryer, L. (1982) Purification and characterization of the Γ subunit of the cyclic GMP phosphodiesterase from retinal rod outer segments. J. Biol. Chem. 257:11094–11099

Kühn, H. (1984) Interactions between photoexcited rhodopsin and light-activated enzymes in rods. In: Osborne, N. & Chader, J. (eds.) Progress in retinal research, Pergamon, New York, pp. 123–156

Kühn, H., Bennett, N., Michel-Villaz, M. & Chabre, M. (1981) Interactions between photoexcited rhodopsin and GTP binding protein kinetics and stoichiometric analyses from light-scattering changes. Proc. Natl. Acad. Sci. USA 78:6873–6877

Miki, N., Baraban, J.M., Keirns, J.J., Boyce, J.J. & Bitensky, M.W. (1975) Purification and properties of the light-activated cyclic nucleotide phosphodiesterase of rod outer segments. J. Biol Chem. 250:6320–6327

Sitarramaya, A., Casedevall, C., Bennett, N. & Hakki, S.I. (1988) Contribution of the guanosinetriphosphatase activity of G-protein to termination of light-activated guanosine cyclic 3′,5′-phosphate hydrolysis in retinal rod outer segments. Biochemistry 27:4880–4887

Wensel, T.G. & Stryer, L. (1986) Reciprocal control of retinal rod cyclic GMP phosphodiesterase by its Γ subunit and transducin. Proteins Struct. Funct. Genet. 1:90–99

Wilden, U., Hall, S.W. & Kühn, H. (1986) Phosphodiesterase activation by photoexcited rhodopsin is quenched when rhodopsin is phosphorylated and binds the intrinsic 48 kD protein of rod outer segments. Proc. Natl. Acad. Sci. USA 83:1174–1178

Phosphoinositide Metabolism and Visual Signal Transduction

Z. SELINGER[1] and B. MINKE[2]

1 Introduction

A central question in vision research is the identity of the biochemical pathway that underlies the phototransduction process. It is now well established that cGMP is the internal messenger of visual transduction in vertebrate rods and that a guanine nucleotide-binding protein (the G-protein transducin) couples the photoexcited rhodopsin to cGMP hydrolysis (for review, see Stryer 1986). Much less is known about phototransduction in the invertebrate photoreceptors, and even the identity of the second messenger of phototransduction is still in dispute. This is because both inositol triphosphate ($InsP_3$) (Brown et al. 1984; Fein et al. 1984) and cGMP (Johnson et al. 1986) have been found to excite the *Limulus* ventral photoreceptors. We have initiated combined electrophysiological and biochemical studies of phototransduction in *Drosophila* and *Musca* flies, since the extensive genetic and physiological background of *Drosophila* mutants can be used to assign physiological roles to biochemical reactions that unavoidably must be measured under in vitro conditions that disrupt the spatial arrangement of the photoreceptor cell.

Another advantage of the fly photoreceptors is that their visual photopigment is thermostable and photoreversible with a large spectral difference between the inactive rhodopsin R_{490} and the active metarhodopsin M_{570}. Hence, by applying blue converting light (<490), R is converted to M (80%) and maximally reconverted to R (100%) by red light (>580). Thus, by the use of either blue converting or red regenerating light, the content of metarhodopsin, the active form of the photopigment, can be easily manipulated.

2 Results and Discussion

We have already characterized a guanosine triphosphatase in *Musca* eye membranes whose activity is regulated by rhodopsin to metarhodopsin conversion, suggesting that phototransduction in *Musca* is mediated by a guanine nucleotide-binding protein (G-protein) (Blumenfeld et al. 1985). The present studies were undertaken to identify the next step in the phototransduction cascade, namely, the target for the G-protein action. *Musca* eye membranes, prelabeled in intact cells with [^{3}H]inositol, respond to illumination by an increase in the accumulation of inositol triphosphate ($InsP_3$),

[1]Departments of Biological Chemistry and [2]Physiology, The Hebrew University of Jerusalem, Jerusalem 91904, Israel

40. Colloquium Mosbach 1989
Molecular Mechanisms of Hormone Action

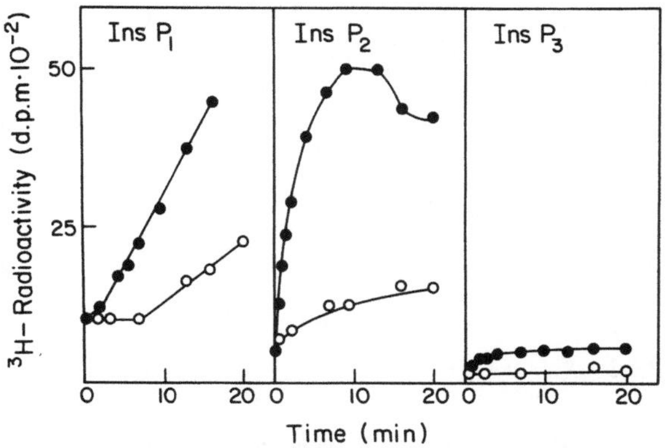

Fig. 1. Light-induced hydrolysis of phosphoinositides in *Musca* eye membranes. *Musca* eyes of a white-eyed mutant were cut and halved with a razor blade. Equivalents of 100 eyes were incubated in the dark for 4 h at 30 °C in 0.5 ml medium containing (in mM) 150 NaCl, 10 KCl, 2 $CaCl_2$, 2 $MgCl_2$, 10 HEPES buffer, pH 7.4, 10 glucose, 5 succinate, and 10 μCi ml^{-1} [^{3}H]inositol. Throughout the incubation the medium was equilibrated with 100% O_2 and gently shaken. At the end of the incubation period, free [^{3}H]inositol was removed by three medium replacements interspaced by 5 min incubation periods. Eye membranes were prepared as previously described (Blumenfeld et al. 1985), except that the homogenization medium contained 1 mM EGTA; the crude membrane fraction was obtained by centrifugation for 15 min at 10 000 × g. Membranes (2 mg ml^{-1}) in homogenization buffer were kept in liquid nitrogen until use. Light-dependent phosphoinositide hydrolysis was measured in medium containing (in mM) 50 MOPS buffer, pH 6.7, 6 $MgCl_2$, 1 DTT, 0.01 GTP, 1 ATP, 0.2 mg ml^{-1} bovine serum albumin, and creatine phosphate and creatine kinase, 5 mM and 50 units ml^{-1}, respectively. The medium was calibrated with Ca-EGTA to give 50 nM free Ca^{2+}. The reaction was initiated by the addition of membranes (500 μg protein ml^{-1}). *Filled circles* represent systems illuminated with blue light (< 490 nM) and *open circles* represent systems incubated in the dark. At each time point, an aliquot of 0.1 ml was removed, the reaction stopped with an equivalent volume of 5% cold trichloracetic acid, centrifuged, and the supernatant analyzed for inositol phosphates by anion exchange chromatography on Dowex 1 columns. (Berridge et al. 1983)

inositol biphosphate ($InsP_2$), and inositol phosphate, the products of phosphoinositide hydrolysis by phospholipase C-type enzyme (Fig. 1).

In the illuminated system, there is a rapid, small increase in the concentration of $InsP_3$, which levels off after 2 min. The greatest effect of light is seen, however, on the accumulation of $InsP_2$, which increases linearly for the first 10 min, then levels off and slightly decreases at longer incubation periods. The accumulation of $InsP_1$ is characterized by a lag period, indicating that this product may arise by stepwise enzymatic dephosphorylation of polyphosphoinositols. Although not directly tested in cell-free membrane preparations, a moderate, light-induced increase in $InsP_3$ has been found in the *Limulus* (Brown et al. 1984) and squid (Szuts et al. 1985) eyes, and a decreased incorporation of [^{32}P] into phosphatidylinositol 4,5 biphosphate has been observed in the *Limulus* (Brown et al. 1984) and octopus (Yoshioka et al. 1983) eyes, as well as in squid photoreceptor membranes (Vanderberg and Montal 1984). In the *Limulus*, however, both Ins 1,4,5P_3 and Ins 1,4P_2 were found to produce effects similar to those caused by light, and we have also confirmed this observation in *Musca* photoreceptor cells (not shown). Taken together with the much larger light-induced increase in $InsP_2$ as compared to $InsP_3$, which we have observed with the *Musca* eye membrane preparation, the question arises whether $InsP_2$ or $InsP_3$ is the internal transmitter of

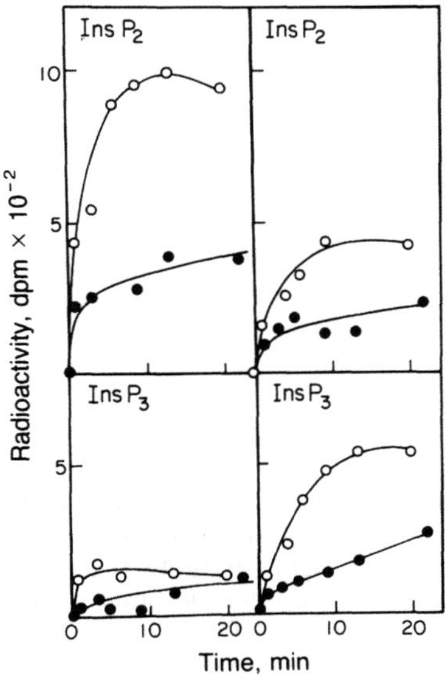

Fig. 2. Effect of 2,3 diphosphoglycerate on light-induced phosphoinositide hydrolysis in *Musca* eye membranes. Incubation conditions were as described in Fig. 1. Where indicated, 2,3 diphosphoglycerate + $MgCl_2$, 10 mM each were added. *Filled circles* represent systems incubated in the dark; *open circles* represent systems illuminated with blue light (< 490)

phototransduction in *Musca* photoreceptor cells. This question was tested directly by the addition of 2,3 diphosphoglycerate (DPG), a known inhibitor of $InsP_3$ phosphatase (Downes et al. 1982). In the presence of 10 mM DPG, there is a substantial decrease in the accumulation of $InsP_2$ and a concurrent large increase in the accumulation of $InsP_3$, which now becomes the major product of the light-induced phosphoinositide hydrolysis (Fig. 2).

The preference of the light-dependent phospholipase C enzyme for phosphatidylinositol 4,5 biphosphate (PIP_2) over phosphatidylinositol 4 phosphate (PIP) is further strengthened by our finding that under our assay conditions (in the presence of ATP and regeneration system), the concentration of PIP in the fly membranes is five times higher than the concentration of PIP_2, the immediate substrate for $InsP_3$ production. Yet when the hydrolysis of $InsP_3$ is blocked by DPG, more $InsP_3$ is accumulated than $InsP_2$. It is also evident that *Musca* eye membranes are endowed with the necessary enzymatic system to eliminate $InsP_3$ after it has been produced. Both light-dependent preferential production of $InsP_3$ and the presence of a turn-off mechanism to stop its action are consistent with an internal transmitter role for $InsP_3$ in fly phototransduction.

We have previously observed that in *Musca* eye membranes, blue light caused an increased GTPase activity both during and long after the light has been turned off. In contrast, red light increases the GTPase activity only during illumination. These findings reflect, in the test tube, the prolonged depolarization after-potential (PDA) observed in the intact fly photoreceptor cell (Blumenfeld et al. 1985). This phenomenon has been attributed to the conversion of neighboring photopigment molecules to metarhodopsin, which thus become resistant to inactivation (Hillman et al. 1983). By monitoring the accumulation of $InsP_2$, it was determined that the PDA phenomenology is also expressed in light-dependent phosphoinositide hydrolysis (Fig. 3).

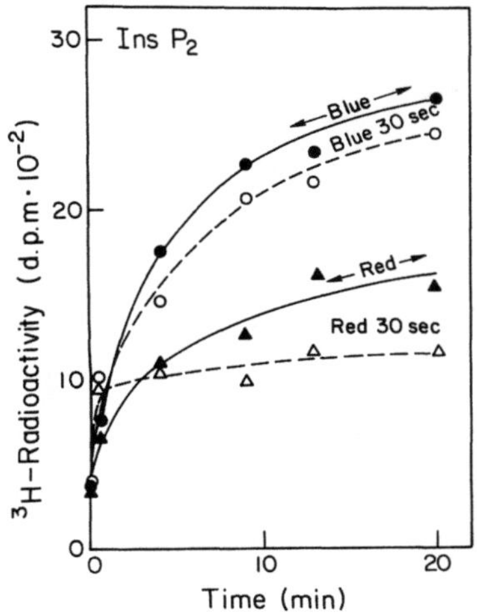

Fig. 3. Effect of illumination with light of different wavelengths on phosphoinositide hydrolysis in *Musca* eye membranes. Incubation conditions were as described in Fig. 1. The reaction was initiated by illumination with either red (< 580) or blue (< 490) light, as indicated

A brief illumination for 30 s with red light gave a burst of $InsP_2$ production, which immediately leveled off. On the other hand, continuous illumination with red light resulted in the continuous production of $InsP_2$. In contrast, brief illumination with blue light caused continuous production of $InsP_2$, thus further supporting the relevance of light-dependent phosphoinositide hydrolysis to the phototransduction process in the fly photoreceptor cell.

Several transduction mechanisms, including phosphoinositide hydrolysis, are mediated by a guanine nucleotide-binding protein (G-protein) which is active in the GTP-bound form and reversed to the inactive state upon hydrolysis of the bound GTP to GDP (Cassel et al. 1977; Fung and Stryer 1980; Cockroft and Gomperts 1985). Since omitting GTP from the incubation system of *Musca* eye membranes had little, if any, effect, the role of a G-protein in phosphoinositide hydrolysis was tested by the addition of GTPγS or GDPβS hydrolysis-resistant analogs of GTP and GDP, respectively (Eckstein et al. 1979). In the control system incubated in the absence of GTPγS, illumination for 30 s with blue light increased the rate of phosphoinositide hydrolysis for about 10 min. However, after illumination for 30 s with red light, the rate of $InsP_2$ accumulation is not significantly different from that of a control system incubated in the dark. In contrast, in the presence of GTPγS, illumination for 30 s with either blue or red light persistently activated the hydrolysis of phosphoinositides (Fig. 4).

It is noteworthy that accumulation of $InsP_2$, after illumination in the presence of GTPγS, proceeds for a longer period and reaches a greater extent than after blue illumination in the presence of GTP. These results indicate that neither inactivation of the phospholipase C enzyme nor exhaustion of the polyphosphoinositide substrates account for the cessation of phosphoinositide hydrolysis in the presence of GTP (Fig. 4). Conceivably an inactivation mechanism, operating at the photopigment level similar to inactivation of vertebrate rod rhodopsin by rhodopsin kinase, also takes place in the *Musca* eye membrane preparation.

As expected, once GTPγS is bound to the G-protein, phosphoinositide hydrolysis is no longer dependent on the continuous action of metarhodopsin. Further

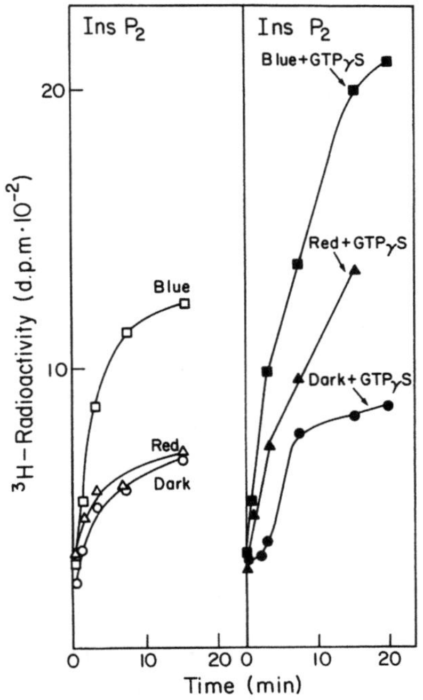

Fig. 4. Stimulation of inositol disphosphate production by GTPγS. Preparation of membranes and incubation conditions were as described in Fig. 1. The systems containing GTPγS (10 μM) did not contain GTP. The reaction was initiated by illumination for 30 s with either red or blue light (>580 and <490, respectively), as indicated

demonstration that light-dependent phosphoinositide hydrolysis is under the control of a G-protein comes from experiments with GDPβS, which unlike GDP is resistant to enzymatic hydrolysis as well as to phosphorylation by nucleoside-diphosphate kinase (Eckstein et al. 1979). In the presence of GDPβS, the rate of the light-induced accumulation of $InsP_2$ is equal to that of a system incubated in the dark, thus indicating that phosphoinositide hydrolysis in *Musca* eye membranes is under the stringent control of the G-protein (not shown). In accordance with these results, a blockade of visual excitation and adaptation by GDPβS has been recently demonstrated in *Limulus* photoreceptor cells (Fein 1986).

Two of the criteria of an internal second messenger are: (1) that introduction of the putative messenger into the cell will reproduce the physiological response, and (2) agents which inhibit the inactivation of the messenger will potentiate the effect of the physiological stimulus. These criteria were tested by introducing $InsP_3$ and DPG into *Musca* photoreceptor cells. The introduction of $InsP_3$ + DPG (not shown) caused a partial depolarization for an extended period, and the elevated baseline had much larger voltage fluctuations (noise).

The visual *Drosophila* mutant *norph A*, which in some alleles completely fails to produce receptor potential in response to light, is particularly useful to test the involvement of $InsP_3$ in photoexcitation. The *norph A* mutant has been found previously to have decreased diglyceride kinase and phospholipase C activities (Yoshioka et al. 1983; Inoue et al. 1985). It is not clear from these studies, however, whether these enzymes are light-dependent and which of the two enzymes is the primary cause of the mutant phenotype. A decreased diglyceride kinase activity can give rise to persistent activation of protein kinase C, which in turn may cause secondary changes in phosphoinositide metabolism.

To overcome these difficulties, we have employed the temperature-sensitive allele *norph A* ^{H52}. This mutant has normal receptor potential at the permissive temperatures of 19–30 °C, but promptly loses its receptor potential when shifted to the restrictive temperature of >34 °C (Deland and Pak 1973). It should be pointed out that *norph A* ^{H52} differs from other *norph A* alleles that are developmentally temperature-sensitive (Inoue et al. 1985). These mutants reveal phenotype (absence of light coincidence receptor potential) only if grown at the restrictive temperature throughout their life cycle, whereas *norph A* ^{H52} retains its temperature sensitivity even when grown at permissive temperatures. Light-dependent phosphoinositide hydrolysis was measured by monitoring the accumulation of $InsP_2$ in reaction mixtures containing either *norph A* ^{H52} or wild-type derived head membranes that were prelabeled with [^{3}H]inositol at the permissive temperature of 21 °C. In the systems containing wild-type membranes, there was practically no difference in phosphoinositide hydrolysis at 21 ° or 37 °C. On the other hand, incubation of *norph A* ^{H52} membranes at the restrictive temperature of 37 °C considerably reduced the rate and extent of $InsP_2$ accumulation, as compared with the system that was incubated at 21 °C. This effect is greatly enhanced in *norph A* ^{H52} membranes that had been preincubated for 4 min at the restrictive temperature of 37 °C, but not in preincubated wild-type membranes or in mutant membranes that had been preincubated at 21 °C (Fig. 5). Since the light-dependent phospholipase C activity is coupled to photoexcited rhodopsin by a

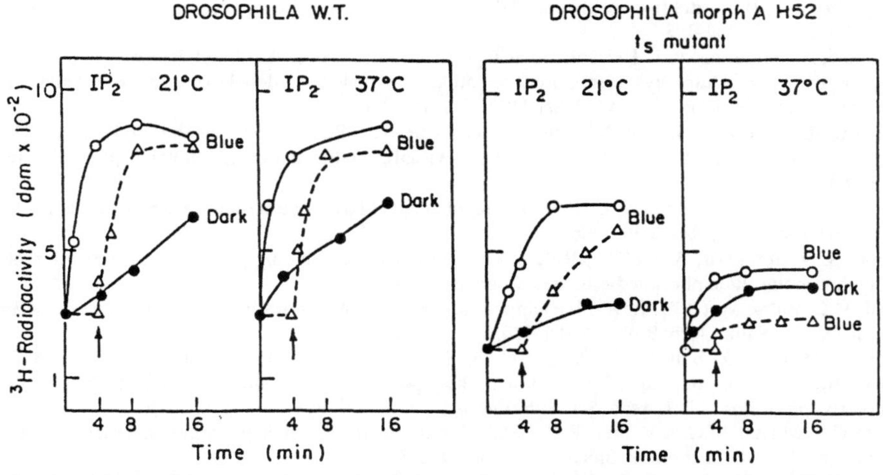

Fig. 5. Light-dependent phosphoinositide hydrolysis by normal (*W.T.*) *Drosophila* and by *norph A* ^{H52} mutant eye membranes incubated at permissive and restrictive temperatures. *Drosophila* (normal and *norph A* ^{H52} temperature-sensitive mutant) flies up to 5-days-old were used. Incorporation of [^{3}H]inositol and preparation of crude *Drosophila* eye membranes were carried out as described previously (Devary et al. 1987). The reaction was initiated by addition of membranes to reaction media preequilibrated at the indicated temperatures and processed as described in Fig. 1. Systems depicted by *dotted lines* were preincubated for 4 min at the indicated temperatures in medium lacking ATP and regeneration system. These latter components were subsequently added (*arrow*) to initiate the reaction

G-protein, and diglyceride kinase activity was reported to be greatly reduced in *norph A* mutants (Yoshioka et al. 1983), we have separately assayed the G-protein by measuring its light-dependent GTPase activity and the diglyceride kinase by measuring incorporation of labeled phosphate from [γ-^{32}P]ATP to exogenously added diglyceride. None of these activities were diminished by a temperature shift from 21° to 37°C (not shown).

The electrophysiological biochemical and genetic evidence that has accumulated over the last few years suggests that vertebrates and invertebrates have evolved profoundly different phototransduction mechanisms in their photoreceptors. It is now clear that in contrast to vertebrate photoreceptors that evolved a highly specific cGMP-mediated transduction mechanism unique to these cells, the invertebrate photoreceptors use the ubiquitous inositol lipid signaling system, a mechanism that operates in almost every eukaryotic cell (Berridge 1987). It is therefore likely that the information gained from studies of phototransduction in the fly photoreceptor will be widely relevant to a great number of cells and a variety of biological phenomena.

Acknowledgments. This work was supported by grants from the European Community and from the National Institutes of Health, USA.

References

Berridge, M.J., Dawson, M.C., Downes, C.P., Heslop, J.P. & Irvine, R.F. (1983) Changes in the levels of inositol phosphates after agonist-dependent hydrolysis of membrane phosphoinositides. Biochem. J. 212:473–482

Berridge, M.J. (1987) Inositol trisphosphate and diacylglycerol: two interacting second messengers. Annu. Rev. Biochem. 56:159–193

Blumenfeld, A., Erusalimsky, J., Heichal, O., Selinger, Z. & Minke, B. (1985) Light-activated guanosine-triphosphatase in *Musca* eye membranes resembles the prolonged depolarizing after-potential in photoreceptor cells. Proc. Natl. Acad. Sci. USA 82:7116–7120

Brown, J.E., Rubin, L.J., Ghalayini, A.J., Tarver, A.P., Irvine, R.F., Berridge, M.J. & Anderson, R.E. (1984) Myo-inositol polyphosphate may be a messenger for visual excitation in *Limulus* photoreceptors. Nature 311:160–163

Cassel, D., Levkovitz, H. & Selinger, Z. (1977) The regulatory GTPase cycle of turkey erythrocyte adenylate cyclase. J. Cycl. Nucl. Res. 3:393–406

Cockroft, S. & Gomperts, B.D. (1985) Role of guanine nucleotide binding protein in the activation of polyphosphoinositide phosphodiesterase. Nature 314:534–536

Deland, M.C. & Pak, W. (1973) Reversible temperature sensitive phototransduction mutant of *Drosophila melanogaster*. Nature New Biol. 244:184–186

Devary, O., Heichal, O., Blumenfeld, A., Cassel, D., Suss, E., Barash, S., Rubinstein, C.T., Minke, B. & Selinger, Z. (1987) Coupling of photoexcited rhodopsin to inositol phospholipid hydrolysis in fly photoreceptors. Proc. Natl. Acad. Sci. 84:6939–6943

Downes, C.P., Mussat, M.C. & Michell, R.H. (1982) The inositol triphosphate phosphomonoesterase of the human erythrocyte membrane. Biochem. J. 203:169–177

Eckstein, F., Cassel, D., Levkovitz, H., Lowe, M. & Selinger, Z. (1979) Guanosine 5′-0-(2-thiodiphosphate): an inhibitor of adenylate cyclase stimulation by guanine nucleotides and fluoride ions. J. Biol. Chem. 254:9829–9834

Fein, A. (1986) Blockade of visual excitation and adaptation in *Limulus* photoreceptor by GDPβS. Science 232:1543–1545

Fein, A., Payne, R., Corson, D.W., Berridge, M.J. & Irvine, R. (1984) Photoreceptor excitation and adaptation by inositol 1,4,5-triphosphate. Nature 311:157–160

Fung, B.K.K. & Stryer, L. (1980) Photolysed rhodopsin catalyzes the exchange of GTP for GDP in rod outer segment membranes. Proc. Natl. Acad. Sci. USA 77:2500–2504

Hillman, P., Hochstein, S. & Minke, B. (1983) Transduction in invertebrate photoreceptors: the role of pigment bistability. Physiol. Rev. 63:668–772

Inoue, H., Yoskioka, T. & Hotta, Y. (1985) A genetic study of inositol trisphosphate involvement in phototransduction using *Drosophila* mutants. Biochem. Biophys. Res. Commun. 132:512–517

Johnson, E.C., Robinson, P.R. & Lisman, J.E. (1986) Cyclic GMP is involved in the excitation of invertebrate photoreceptors. Nature 324:468–470

Stryer, L. (1986) Cyclic GMP cascade of vision. Ann. Rev. Neurosci. 9:87–119

Szuts, E., Reid, M., Payne, R., Corson, D.W. & Fein, A. (1985) Biochemical and physiological evidence for the involvement of inositol 1,4,5 triphosphate in visual transduction. Biophys. J. 47:202a

Vanderberg, C.A. & Montal, M. (1984) Light-regulated phosphorylation of rhodopsin and phosphoinositides in squid photoreceptor membranes. Biochemistry 23:2347–2352

Yoshioka, T., Inoue, H., Takagi, M., Hayashi, F. & Amakawa, T. (1983) The effect of isobutylmethyl-xanthin on the photoresponse and phosphorylation of phosphatidylinositol in octopus retina. Biochim. Biophys. Acta 755:50–55

Hormonal Regulation of Phospholipases

J.H. EXTON[1]

1 Introduction

It is now widely accepted that many hormones, neurotransmitters, and related agonists exert some of their biological effects by breaking down phosphatidylinositol 4,5-biphosphate (PIP_2) a minor phospholipid in the plasma membrane of their target cells. This produces two signaling molecules: inositol trisphosphate (IP_3) which releases intracellular Ca^{2+}, and diacylglycerol (DAG) which activates protein kinase C (Berridge 1987). There is much evidence that the phospholipase C that catalyzes the breakdown of PIP_2 is regulated by a guanine nucleotide-binding regulatory protein (G-protein) which is activated when the relevant plasma membrane receptors are occupied by agonists (Exton 1988a).

Recent evidence has indicated that many of the agonists that stimulate PIP_2 hydrolysis also cause the breakdown of phosphatidylcholine (PC) to yield DAG, phosphatidic acid (PA) as well as choline and phosphocholine (Exton 1988b). Although the physiological significance of this response is not clear at present, it is observed with many agonists in many tissues and cell types (Table 1). In the following account, I describe our investigations into the molecular mechanisms involved in agonist-stimulated PIP_2 and PC breakdown using the liver as a test tissue.

2 Results

2.1 *Phosphatidylinositol 4,5-Biphosphate Breakdown*

Figure 1 shows the changes in inositol phospholipids induced by the Ca^{2+}-mobilizing agonist vasopressin in isolated rat hepatocytes (Augert et al. 1989a). There is a rapid, large decrease in the content of PIP_2, whereas phosphatidylinositol 4-phosphate (PIP) does not change and phosphatidylinositol (PI) decreases much more slowly. The rapid breakdown of PIP_2 is associated with a parallel increase in IP_3 (Exton 1988a), indicating the activation of a phospholipase C. Experiments utilizing isolated rat liver plasma membranes have shown the activation of a PIP_2 phospholipase C by GTP analogs and by agonists in the presence of these analogs (Uhing et al. 1986; Taylor and Exton 1987). This activation is observed with submicromolar concentrations of GTP analogs, requires millimolar Mg^{2+}, and is competitively inhibited by a stable analog of GDP

[1]Howard Hughes Medical Institute and Departments of Molecular Physiology and Biophysics, and Pharmacology, Vanderbilt University School of Medicine, Nashville, TN 37232, USA

40. Colloquium Mosbach 1989
Molecular Mechanisms of Hormone Action

Table 1. Agonist Stimulation of Phosphatidylcholine Breakdown in Different Cell Types

Cell Type	Agonists
Hepatocyte	Vasopressin, angiotensin II, epinephrine (α_1), ATP (P_2), phorbol ester
Hepatoma (HepG2)	Serum factor(s), phorbol ester
Kidney cell (MDCK-D1)	Epinephrine (α_1), phorbol ester
Promyelocyte (HL-60)	f-met-leu-phe, phorbol ester, diacylglycerol
Neuroblastoma (N4TG1 NG108-15)	Serum factor(s), phorbol ester
Pheochromocytoma (PC-12)	Serum factor(s), phorbol ester
Pre-adipocyte (3T3-L1)	PDGF, phorbol ester
Fibroblast (Swiss 3T3, 3T3-A31)	Bombesin, phorbol ester
Aortic smooth muscle cell (A10)	Vasopressin
Rat embryo cell (REF52)	Vasopressin
T-lymphocyte (Jurkat)	Interleukin 1
Neutrophil	f-met-leu-phe, phorbol ester
Endothelial cell	Thrombin, bradykinin
Heart	Acetylcholine
Brain	Acetylcholine
Pancreas	Acetylcholine
Platelet	Thrombin
Myoblast	Phorbol ester
HeLa cell	Phorbol ester
Uterine decidua cell	Phorbol ester
Pituitary (GH3)	Diacylglycerol
Mast cell	Antigen

indicating the involvement of a G-protein (Uhing et al. 1986; Taylor and Exton 1987). Evidence that Ca^{2+}-mobilizing receptors are linked to a G-protein(s) is provided by the effects of GTP and its analogs on the binding of agonists to these receptors (Lynch et al. 1986) and by the stimulation of a low K_m GTPase activity in liver plasma membranes by Ca^{2+}-mobilizing agonists (Fitzgerald et al. 1986).

Recent work has shown that most tissues contain several phosphoinositide phospholipase C activities (Exton 1988a). We have purified the phospholipase C responsible for G-protein-stimulated PIP_2 hydrolysis in liver membranes and have shown that it has an M_r of 148 000 and is identical immunologically to the type II phospholipase C purified from bovine brain by Ryu et al. (1987a,b). However, the specific G-protein(s) involved in the activation of the liver phospholipase remains undefined, although it is clear that it is not a substrate for either pertussis or cholera toxin (Uhing et al. 1986). Rat liver plasma membranes contain large amounts of a toxin-insensitive G-protein with a 40 kDa α-subunit, but attempts to reconstitute this with the phospholipase have been unsuccessful.

2.2 Phosphatidylcholine Breakdown

The first piece of evidence that we obtained indicating that Ca^{2+}-mobilizing agonists stimulate the breakdown of another phospholipid besides PIP_2 came from HPLC analyses of the DAG accumulating in response to vasopressin in rat liver cells (Bocckino et al. 1985). Fatty acid analysis of the DAG indicated that it probably came

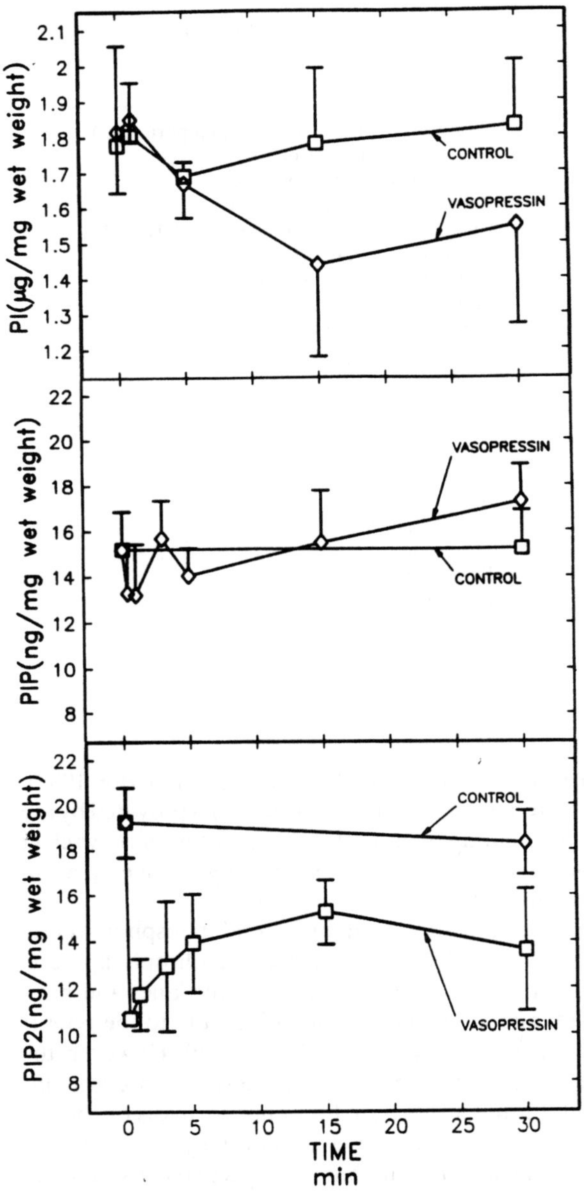

Fig. 1. Changes in inositol phospholipids induced by vasopressin in rat hepatocytes. Hepatocytes were incubated with saline (control) or 100 nM vasopressin for the indicated times. Incubations were stopped in ice-cold $CHCl_3$/MeOH and the phosphoinositides extracted and analyzed as described by Augert et al. (1989a). From Augert et al. (1989a) by permission of the authors and publisher

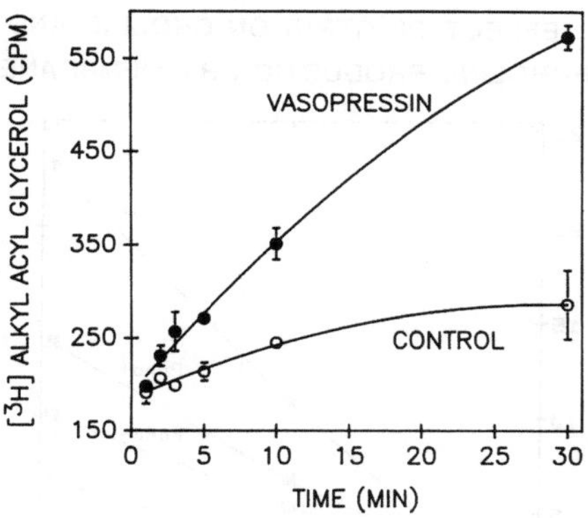

Fig. 2. Vasopressin stimulation of [^{3}H]alkyl-PC to [^{3}H]alkyl-acylglycerol in rat hepatocytes. Hepatocytes were incubated for 30 min with 1-*O*-[^{3}H]alkyl-2-lyso-*sn*-glycero-3-phosphocholine to label the PC and then incubated with saline (control) or 100 nM vasopressin for the times shown. [^{3}H]alkyl-acylglycerol was isolated and counted as described by Augert et al. (1989b)

from PC. More refined HPLC analyses of the molecular species of DAG (Augert et al. 1989b) and chemical measurements of the changes in inositol phospholipids and DAG (Bocckino et al. 1985, 1987; Augert et al. 1989a) have confirmed this conclusion. More direct proof that certain agonists promote the breakdown of PC to DAG in hepatocytes is presented in Fig. 2. This shows that in experiments in which the PC of hepatocytes was labeled for 30 min with [^{3}H]alkyl-lyso-glycerophosphocholine, vasopressin stimulated the production of [^{3}H]alkyl-acylglycerol. Similar data were obtained when PC was labeled with [^{14}C]lyso-PC and [^{14}C]DAG formation was measured (Augert et al. 1989b). There was also a stimulation of the release of [^{3}H]choline and [^{3}H]phosphocholine when PC was labeled by incubating the cells with [^{3}H]choline for 90 min. In hepatocytes and other cell types there is now evidence that PC is broken down to PA by a phospholipase D activity as well as to DAG by a phospholipase C activity (Exton 1988b; Slivka et al. 1988; Cabot et al. 1988a; Pai et al. 1988a,b; Martin and Michaelis 1988; Agwu et al. 1989).

2.3 Mechanisms of Hormonal Stimulation of Phosphatidylcholine Breakdown

Evidence for three mechanisms of hormonal stimulation of PC breakdown has been presented. The first is G-protein-mediated stimulation of PC breakdown by phospholipase C and D activities. The second is protein kinase C-mediated activation of these phospholipases. The third is Ca^{2+} stimulation of the phospholipases.

The evidence for control of PC breakdown through a G-protein(s) comes from studies with isolated rat liver plasma membranes. As illustrated in Fig. 3, addition of GTPγS to these membranes stimulates the production of DAG, PA, choline, and phosphocholine (Bocckino et al. 1987; Irving and Exton 1987). Since there is an associated decrease in PC, these data are consistent with the activation of PC breakdown by phospholipase C and D. The effect is observed with micromolar or submicromolar concentrations of GTP analogs and is inhibited by a stable GDP analog (Irving and Exton 1987). Purinergic agonists (ATP and ADP) also stimulate PC breakdown in liver plasma membranes, but only in the presence of a GTP analog (Irving and Exton 1987; Bocckino et al. 1987). Neither cholera nor pertussis toxin alters the hydrolysis of PC to choline and phosphocholine (Irving and Exton 1987) indicating that the putative G-protein involved is not a substrate for these toxins.

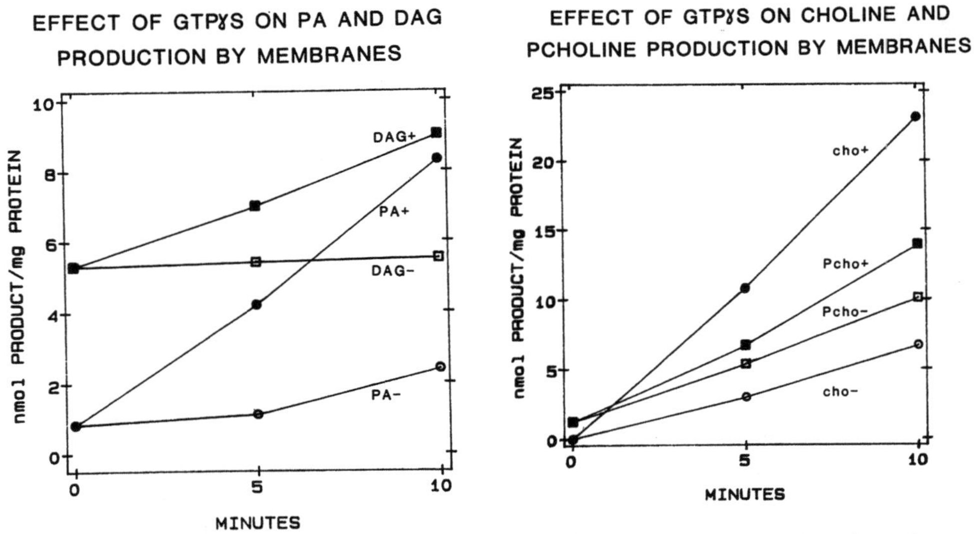

Fig. 3. Stimulation of PA, DAG, choline, and phosphocholine production in rat liver plasma membranes by GTPγS. Rat liver plasma membranes were incubated with 20 μM GTPγS for 0, 5, or 10 min and the levels of PA, DAG, choline, and phosphocholine measured as described by Bocckino et al. (1987). From Bocckino et al. (1987) by permission of the authors and publisher

Evidence that protein kinase C controls PC breakdown comes from studies in which tumor promoting phorbol esters and synthetic DAGs promote the release of choline and phosphocholine from cells or stimulate the breakdown of [^{3}H]alkyl-PC to labeled DAG and PA (Mufson et al. 1981; Guy and Murray 1982; Daniel et al. 1986; Besterman et al. 1986; Muir and Murray 1987; Schrey et al. 1987; Kolesnick and Paley 1987; Liscovitch et al. 1987; Takuwa et al. 1987; Cabot et al. 1988a; Agwu et al. 1989; Augert et al. 1989b). Since all the agonists that elicit PC breakdown also promote rapid PIP_2 hydrolysis, it is possible that the initial increase in DAG arising from PIP_2 triggers PC breakdown through activation of protein kinase C. Additionally, PC hydrolysis due to activation of protein kinase C might represent a positive feedback to ensure elevated levels of DAG for prolonged physiological effects.

Several studies have indicated that an elevation in cytosolic Ca^{2+} can induce PC breakdown. For example, the Ca^{2+} ionophore A23187 elicits the conversion of [^{3}H]alkyl-PC to [^{3}H]alkyl-acyl-glycerol or [^{3}H]alkyl-PA in hepatocytes and neutrophils (Augert et al. 1989b; Agwu et al. 1989). Furthermore, the effects of agonists on DAG formation or [^{3}H]alkyl-PC breakdown in several cell types are partly dependent upon the presence of Ca^{2+} in the medium (Bocckino et al. 1985; Pickford et al. 1987; Pai et al. 1988b; Polverino and Barritt 1988; Augert et al. 1989b), and there is evidence that an increase in cytosolic Ca^{2+} in polymorphonuclear leukocytes can induce the formation of DAG from a phospholipid that does not contain inositol (Truett et al. 1988). These observations raise the possibility that part of the stimulatory action of Ca^{2+}-mobilizing agonists in PC breakdown is due to the stimulation of phospholipases by the increase in cytosolic Ca^{2+}.

3 Discussion

Although there is abundant evidence that Ca^{2+}-mobilizing agonists control PIP_2 phospholipase C activity via G-proteins, the nature of these proteins remains undefined. It is clear that more than one G-protein is involved since in some cell types (neutrophils, mast cells, mesangial cells) the response is blocked by pertussis toxin, whereas in others (hepatocytes, adrenal, thyroid, and pituitary cells) it is not (Exton 1988a). The inability to identify the relevant G-proteins is due to the fact that attempts to reconstitute purified G-proteins from various sources with purified phospholipases have been unsuccessful. In addition, there have been no reports of the successful expression or introduction of G-proteins into cells to restore or amplify agonist effects on PIP_2 breakdown. It is possible that some of the G-proteins involved in this signal transduction have been already identified by protein purification or cloning techniques, but until their functions are defined this remains speculation. In the absence of any functional coupling of G-proteins to the phospholipase C, the mechanism by which activation of this enzyme occurs also remains uncertain.

An interesting new finding is the discovery that many hormones, neurotransmitters, and growth factors stimulate PC breakdown in addition to PIP_2 hydrolysis. As shown in Table 1, this response is very widespread. An intriguing discovery is that it involves both phospholipases C and D, with the production of both DAG and PA. The mechanisms by which the various agonists activate these phospholipases are relatively undefined. Evidence has been obtained for mechanisms involving G-proteins, protein kinase C, and Ca^{2+}.

Another point that requires definition is the physiological significance of agonist-stimulated PC breakdown in cells. One proposal that is receiving increasing support is that PC contributes more DAG for activation of protein kinase C than does PIP_2 (Augert et al. 1989a). Another surprising finding is that large amounts of PA are produced from PC breakdown at early times of agonist action when DAG accumulation is not detectable, indicating that phospholipase D is activated before phospholipase C (Bocckino et al. 1987; Pai et al. 1988a,b; Cabot et al. 1988a; Agwu et al. 1989). This has led Cabot et al. (1988b) and Martin (1988) to postulate that some of the DAG that accumulates is derived from PA through the action of phosphatidate phosphohydrolase. The physiological functions of agonist-generated PA remain unknown, although it has been proposed to function as a mitogen (Moolenaar et al. 1986; Yu et al. 1988). Its early formation suggests that it has a role in more rapid responses, but these remain unknown.

4 Conclusions

1. Although it is generally accepted that many hormones and neurotransmitters regulate inositol phospholipid breakdown through the mediation of G-proteins, the nature of these proteins and the molecular mechanisms involved still remain unknown.
2. Most of the agonists that stimulate inositol phospholipid hydrolysis also promote phosphatidylcholine breakdown through activation of phospholipase C and D activities.

3. There is evidence that agonist-stimulated phosphatidylcholine breakdown involves G-proteins, protein kinase C, and Ca^{2+} ions, but the mechanisms are not yet defined.

References

Agwu, D.E., McPhail, L.D., Chabot, M.C., Daniel, L.W., Wykle, R.L. & McCall, C.E. (1989) Choline-linked phosphogylcerides: a source of phosphatidic acid and diglycerides in stimulated neutrophils. J. Biol. Chem. 264:1405–1413

Augert, G., Blackmore, P.F. & Exton, J.H. (1989a) Changes in the concentration and fatty acid composition of phosphoinositides induced by hormones in hepatocytes. J. Biol. Chem. 264:2574–2580

Augert, G., Bocckino, S.B., Blackmore, P.F. & Exton, J.H. (1989b) Hormonal stimulation of diacylglycerol formation in hepatocytes. Evidence for phosphatidylcholine breakdown. J. Biol. Chem. (in press)

Berridge, M.J. (1987) Inositol trisphosphate and diacylglycerol: two interacting second messengers. Annu. Rev. Biochem. 56:159–193

Besterman, J.M., Duronio, V. & Cuatrecasas, P. (1986) Rapid formation of diacylglycerol from phosphatidylcholine: a pathway for generation of a second messenger. Proc. Natl. Acad. Sci. USA 83:6785–6789

Bocckino, S.B., Blackmore, P.F. & Exton, J.H. (1985) Stimulation of 1,2-diacylglycerol accumulation in hepatocytes by vasopressin, epinephrine and angiotensin II. J. Biol. Chem. 260:14201–14207

Bocckino, S.B., Blackmore, P.F., Wilson, P.B. & Exton, J.H. (1987) Phosphatidate accumulation in hormone-treated hepatocytes via a phospholipase D mechanism. J. Biol. Chem. 262:15309–15315

Cabot, M.C., Welsh, C.J., Zhang, Z.-C., Cao, H.-T., Chabbott, H. & Lebowitz, M. (1988a) Vasopressin, phorbol diesters and serum elicit choline glycerophospholipid hydrolysis and diacylglycerol formation in nontransformed cells: transformed derivatives do not respond. Biochim. Biophys. Acta 959:46–57

Cabot, M.C., Welsh, C.J., Cao, H.-T. & Chabbott, H. (1988b) The phosphatidylcholine pathway of diacylglycerol formation stimulated by phorbol diesters occurs via phospholipase D activation. FEBS Lett. 233:153–157

Daniel, L.W., Waite, M. & Wykle, R.L. (1986) A novel mechanism of diglyceride formation: 12-0-tetradecanoylphorbol-13-acetate stimulates the cyclic breakdown and resynthesis of phosphatidylcholine. J. Biol. Chem. 261:9128–9132

Exton, J.H. (1988a) The roles of calcium and phosphoinositides in the mechanisms of α_1-adrenergic and other agonists. Rev. Physiol. Biochem. Pharmacol. 111:118–224

Exton, J.H. (1988b) Mechanisms of action of calcium-mobilizing agonists: some variations on a young theme. FASEB J. 2:2670–2676

Fitzgerald, T.J., Uhing, R.J. & Exton, J.H. (1986) Solubilization of the vasopressin receptor from rat liver membranes. J. Biol. Chem. 261:16871–16877

Guy, G.R. & Murray, A.W. (1982) Tumor promoter stimulation of phosphatidylcholine turnover in HeLa cells. Cancer Res. 42:1980–1985

Irving, H.R. & Exton, J.H. (1987) Phosphatidylcholine breakdown in rat liver plasma membranes. J. Biol. Chem. 262:3440–3443

Kolesnick, R.N. & Paley, A.E. (1987) 1,2-Diacylglycerols and phorbol esters stimulate phosphatidylcholine metabolism in GH_3 pituitary cells: evidence for separate mechanisms of action. J. Biol. Chem. 262:9204–9210

Liscovitch, M., Blusztajn, J.K., Freese, A. & Wurtman, R.J. (1987) Stimulation of choline release from NG108-15 cells by 12-O-tetradecanoylphorbol 13-acetate. Biochem. J. 241:81–86

Lynch, C.J., Prpic, V., Blackmore, P.F. & Exton, J.H. (1986) Effect of islet-activating pertussis toxin on the binding characteristics of Ca^{2+}-mobilizing hormones and on agonist activation of phosphorylase in hepatocytes. Mol. Pharmacol. 29:196–203

Martin, T.W. (1988) Formation of diacylglycerol by a phospholipase D-phosphatidate phosphatase pathway specific for phosphatidylcholine in endothelial cells. Biochim. Biophys. Acta 962:282–296

Martin, T.W. & Michaelis, K.C. (1988) Bradykinin stimulates phosphodiesteratic cleavage of phosphatidylcholine in cultured endothelial cells. Biochem. Biophys. Res. Commun. 157:1271–1279

Moolenaar, W.H., Kruijer, W., Tilly, B.C., Verlaan, I., Bierman, A.J. & deLaat, S.W. (1986) Growth factor-like action of phosphatidic acid. Nature (London) 310:644–649

Mufson, R.A., Okin, E. & Weinstein, I.B. (1981) Phorbol esters stimulate the rapid release of choline from prelabelled cells. Carcinogenesis 2:1095–1102
Muir, J.G. & Murray, A.W. (1987) Bombesin and phorbol ester stimulate phosphatidylcholine hydrolysis by phospholipase C: evidence for a role of protein kinase C. J. Cell. Physiol. 130:382–391
Pai, J.-K., Siegel, M.I., Egan, R.W. & Billah, M.M. (1988a) Activation of phospholipase D by chemotactic peptide in HL-60 granulocytes. Biochem. Biophys. Res. Commun. 150:355–364
Pai, J.-K., Siegel, M.I., Egan, R.W. & Billah, M.M. (1988b) Phospholipase D catalyzes phospholipid metabolism in chemotactic peptide-stimulated HL-60 granulocytes. J. Biol. Chem. 263:12472–12477
Pickford, L.B., Polverino, A.J. & Barritt, G.J. (1987) Evidence from studies employing radioactively labelled fatty acids that the stimulation of flux through the diacylglycerol pool is an early action of vasopressin on hepatocytes. Biochem. J. (1987) 245:211–216
Polverino, A.J. & Barritt, G.J. (1988) On the source of the vasopressin-induced increases in diacylglycerol in hepatocytes. Biochim. Biophys. Acta 970:75–82
Ryu, S.H., Cho, K.S., Lee, K.-Y., Suh, P.-G. & Rhee, S.G. (1987a) Purification and characterization of two immunologically distinct phosphoinositide-specific phospholipases C from bovine brain. J. Biol. Chem. 262:12511–12518
Ryu, S.H., Suh, P.-G., Cho, K.S., Lee, K.-Y. & Rhee, S.G. (1987b) Bovine brain cytosol contains three immunologically distinct forms of inositolphospholipid-specific phospholipase C. Proc. Natl. Acad Sci. USA 84:6649–6653
Schrey, M.P., Read, A.M. & Steer, P.J. (1987) Stimulation of phospholipid hydrolysis and arachidonic acid mobilization in human uterine decidua cells by phorbol ester. Biochem. J. 246:705–713
Slivka, S.R., Meier, K.E. & Insel, P.A. (1988) α_1-Adrenergic receptors promote phosphatidylcholine hydrolysis in MDCK-D1 cells: a mechanism for rapid activation of protein kinase C. J. Biol. Chem. 263:12242–12246
Takuwa, N., Takuwa, Y. & Rasmussen, H. (1987) A tumour promoter, 12-O-tetradecanoylphorbol 13-acetate, increases cellular 1,2-diacylglycerol content through a mechanism other than phosphoinositide hydrolysis in Swiss-mouse 3T3 fibroblasts. Biochem. J. 243:647–653
Taylor, S.J. & Exton, J.H. (1987) Guanine-nucleotide and hormone regulation of polyphosphoinositide phospholipase C activity of rat liver plasma membranes. Biochem. J. 248:791–799
Truett, A.P., III, Verghese, M.W., Dillon, S.B. & Snyderman, R. (1988) Calcium influx stimulates a second pathway for sustained diacylglycerol production in leukocytes activated by chemoattractants. Proc. Natl. Acad. Sci. USA 85:1549–1553
Uhing, R.J., Prpic, V., Jiang, H. & Exton, J.H. (1986) Hormone-stimulated polyphosphoinositide breakdown in rat liver plasma membranes. J. Biol. Chem. 261:2140–2146
Yu, C.-L., Tsai, M.-H. & Stacey, D.W. (1988) Cellular rat activity and phospholipid metabolism. Cell 52:63–71

Part 3

Regulation of Ion Channels

H. Reuter[1]

1 Introduction

Higher organisms need special systems for signal transmission. They are required for the coordination of cell and organ functions. Ion channels in cell membranes play a crucial role in this information process. They are involved in a multitude of functions, such as synaptic transmission in the nervous systems, conduction of action potentials in excitable cells, intercellular junctional connections in many organs, or regulation of localized intracellular ion concentrations. Many ion channels are directly or indirectly coupled to receptors for neurotransmitters and hormones. This provides for considerable plasticity in the information process. The aim of the present report is to describe some basic properties and modulatory regulations of ion channels. In view of the abundance of information in this field, it was only possible to select a few representative examples.

2 Definition, Classification, and Measurement of Ion Channels

Ion channels are proteins embedded in the lipid bilayers of cell membranes. Under appropriate conditions they constitute pores through which ions can flow down their respective electrochemical gradients. Various types of channels differ in their ion *selectivity*, in their *gating*, and in the way they are *modulated*. Each channel type selects certain ion species that are allowed to pass through the pore when the channel is open and excludes others. For example, Na channels favor Na ions over K ions, while the opposite is true for K channels. The transitions between open and closed conformations of a channel are called gating. The availability and responsiveness of the channels for gating can be modulated by different mechanisms acting either inside or outside the cell. Neurotransmitters and hormones are physiological modulators involved in the functional coordination of different organs.

Ion channels can be broadly classified according to the main energy inputs that are necessary and sufficient for their gating. A change in membrane potential will cause *voltage-gated* ion channels in excitable membranes to open or close (see Table 1 for examples). By contrast, *ligand-gated* ion channels change their conformation when specific molecules bind to a receptor site of the channel protein. They can be subdivided into two classes (see Table 1 for examples): *primary* ligand-gated channels

[1]Pharmakologisches Institut, Universität Bern, Friedbühlstr. 49, CH-3010 Bern, Switzerland

40. Colloquium Mosbach 1989
Molecular Mechanisms of Hormone Action

Table 1. Ion channel classification according to energy inputs required for gating (examples)

Voltage-gated channels:
Na^+ channels
K^+ channels
Ca^{2+} channels
Primary ligand-gated channels:
Acetylcholine-gated channels (nicotinic)
γ-Aminobutyric acid (GABA)-gated channels
Glycine-gated channels
Secondary ligand-gated channels:
Ca^{2+}-gated channels
G-protein-gated channels
cGMP-gated channels
ATP-gated channels

change their ion conductance state by direct binding of a ligand, i.e., without involvement of a second messenger. *Secondary* ligand-gated channels depend on a reaction with a molecule whose availability is due to another cellular event (occupancy of a distant receptor, changes in cell metabolism, etc.). *Modulation* of a channel means that the primary energy input required for channel gating is affected by a secondary event (e.g., a biochemical reaction induced by a second messenger) in such a way that gating is altered. Phosphorylation of a channel protein is a good example for such a modulatory reaction (Levitan 1988).

Since gating of a channel is followed by changes in its ion conductance, electrophysiological techniques, which measure these conductances, are useful tools to investigate the functional state of ion channels. The most important technique to study ionic conductance properties of membrane channels is the patch clamp method (Hamill et al. 1981).

Single channel recordings by means of this technique have greatly advanced our understanding of these membrane proteins. Not only can we measure directly and separately the gating behavior of the various channel types, but we can also assess the modulation of gating that occurs by interference with cell metabolism. For example, one can add ligands, such as neurotransmitters, either into the pipette solution or into the external medium and watch how they alter the gating of channels. The seal between glass pipette and membrane is tight enough to prevent relatively small molecules, such as acetylcholine, from diffusing through it. If channel gating is influenced only after application of the ligands into the pipette solution, this points to a primary ligand-gated channel. If, however, molecules alter channel properties only after bath application, this suggests an effect via second messengers, which either induce (secondary ligand-gated channels) or modulate gating.

3 Modulation of Ion Channels

Ca channels in cardiac cell membranes were the first voltage-gated ion channels shown to be modulated by adrenergic neurotransmitters (for review, see Reuter 1983). Ca channels are of particular interest since they link electrical signals with other cellular events. They are opened during an action potential and contribute to the change in the free Ca ion concentration in the cytoplasm which, in turn, regulates numerous cellular functions. Three distinct types of voltage-dependent Ca channels have been described (T, L, and N; Fig. 1; Nowycky et al. 1985). In the heart only the L-type Ca channel is modulated by catecholamines (Bean 1985). The cascade of events leading to modulation of L-type Ca channel gating is illustrated in Fig. 2. Binding of a β-adrenergic agonist (A) to a β-adrenoceptor (R) leads to activation of the enzyme adenylate cyclase via a GTP-binding protein. The enzyme catalyzes the formation of cAMP from ATP. cAMP binds to the regulatory subunit of protein kinase A and the released catalytic subunit of the enzyme phosphorylates a subunit of the Ca channel. Without being phosphorylated the channel has only a small probability to enter the open state when the membrane potential is depolarized, while the opening probability of the channel is greatly enhanced after phosphorylation (Reuter et al. 1986; Tsien et al. 1986; Kameyama et al. 1986).

A similar type of modulation of ion channels by protein kinase A-dependent phosphorylation seems to be rather widespread. For example, the opening probability of Ca-dependent K channels in neurons from land snails (*Helix roseneri*) is also increased by protein kinase A-dependent phosphorylation (De Peyer et al. 1982; Levitan 1985). In other molluscan neurons, however, the open state probability of K channels is reduced after protein kinase A-dependent phosphorylation. A very interesting example is a K channel in *Aplysia* sensory neurons which is normally open

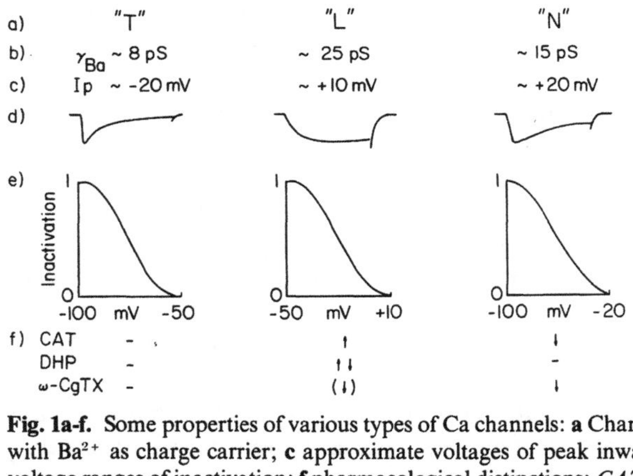

Fig. 1a-f. Some properties of various types of Ca channels: **a** Channel types; **b** single channel conductance with Ba^{2+} as charge carrier; **c** approximate voltages of peak inward Ca currents; **d** current waveform; **e** voltage ranges of inactivation; **f** pharmacological distinctions: *CAT* = catecholamines have little effect (—) on *T* channels, increase (↑) open state probability of *L* channels (via β-adrenoceptor stimulation), and inhibit (↓) current through *N* channels (via α_2-adrenoceptor stimulation); *DHP* = 1,4 dihydropyridines have little effect on *T* and *N* channels. (—), and either increase (↑ activators, agonists) or decrease (↓ blockers, antagonists) open state probability of *L*-type channels. ω-*CgTX* = ω-conotoxin inhibits irreversibly Ca currents through *N*-type and some (but not all) *L*-type channels and has no effect on *T*-type channels (Reuter 1988)

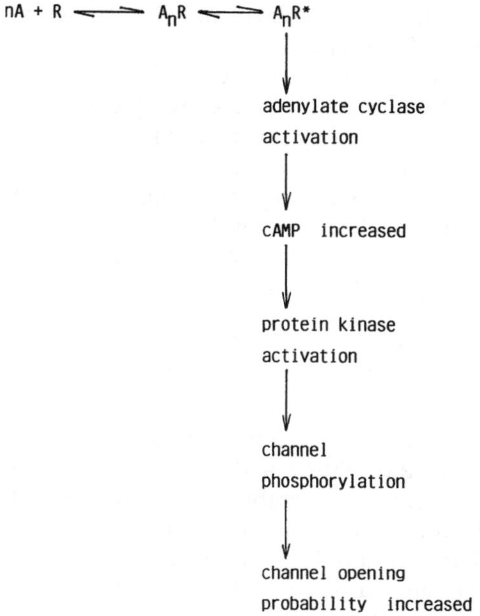

Fig. 2. Cascade of reactions induced by β-adrenoceptor (*R*) agonists (*A*) that finally results in an increased open state probability of a Ca channel. For details, see text

most of the time and closes after protein kinase A-dependent phosphorylation. cAMP is elevated in these cells by the neurotransmitter serotonin. Closure of the K channels produces a prolongation of the action potential duration which, in turn, facilitates neurotransmitter release. This is believed to be related to simple patterns of "learning" by the animal (Belardetti and Siegelbaum 1988; Kandel and Schwartz 1982).

The change in the opening probability of ion channels is only one way by which ions flow through channels can be regulated. Another mechanism is recruitment of new, or previously silent, ion channels by stimulation of protein kinase C. The best example is the unmasking of a certain class of voltage-operated Ca channels in *Aplysia* bag cell neurons. This channel type cannot be recorded in untreated cells but becomes very prominent after exposure to agents that activate protein kinase C (Kaczmarek 1987). This mechanism seems to be very different from the pattern of modulation of L-type Ca channels in heart cells where only the opening probability is increased (see above). In *Aplysia* bag cell neurons, however, a completely new class of previously silent channels is unmasked and, thus, the Ca current into these cells is increased.

By contrast, activation of protein kinase C in *Xenopus* oocytes modulates newly expressed ion channels in different ways. It increases a Ca current, decreases a Na current and $GABA_A$-activated current, and leaves kainate-activated currents unaffected (Sigel and Baur 1988).

In other systems phosphorylation is not an essential step in the regulation of ion channels by nucleotides. ATP gates certain K channels in heart muscle and pancreatic islet cells by direct binding to the channel (Noma 1983; Ashcroft 1988). Similarly, cGMP binds allosterically to cation channels in rod photoreceptors and keeps them open. Light absorbed by rhodopsin causes rapid hydrolysis of cGMP through a cascade of events and, thereby, closes these channels (Fesenko et al. 1985). These are examples of secondary ligand-gated channels (Table 1).

An important new aspect of ion channel modulation is the role played by GTP-binding proteins (G-proteins). These proteins couple surface receptors to ion channels. The first ion channel shown to be regulated by this mechanism was a K channel in heart cells that is coupled via a G-protein to a muscarinic cholinergic receptor (for review, see Neer and Clapham 1988). Opening of K channels by this mechanism slows the heart rate. Meanwhile it has been shown that G-proteins are involved in the gating of many ion channels, notably of voltage-dependent Ca channels (Rosenthal et al. 1988; see also Chaps. 17, 18, this Vol.).

4 Functional Implications of Ion Channel Modulation

In Table 1 only a small selection of many channel types in cell membranes has been listed. Also, the examples of ion channel modulation chosen in this chapter are only a few. However, it should be clear that ion channels in membranes are specific and important structures that are involved in the control of numerous cellular functions. It seems plausible that such control mechanisms should be switched on and off depending on when they are needed. There are various ways by which this can be achieved. One mechanism is by direct binding of regulatory neurotransmitters or hormones to a receptor site on the channel and a subsequent conformational change of the protein. Other mechanisms involve "remote controls" where receptor-regulated activation or inhibition of enzymes results in metabolic products which feed back to ion channels in the membrane. In the heart the importance of ion channel modulation is quite obvious: regulation of frequency and force of the beating heart by neurotransmitters depends on the modulation of the gating of Ca and K channels. The finely tuned interplay between receptor occupancy and channel function also secures the plasticity required for the coordination of cell functions in extremely complex organs, such as the brain.

Acknowledgment. Generous support of the author's work by the Swiss National Science Foundation is gratefully acknowledged.

References

Ashcroft, F.M. (1988) Adenosine 5′-triphosphate-sensitive potassium channels. Annu. Rev. Neurosci. 11:97–118

Bean, B.P. (1985) Two kinds of calcium channels in canine atrial cells. Differences in kinetics, selectivity, and pharmacology. J. Gen. Physiol. 86:1–30

Belardetti, F. & Siegelbaum, S.A. (1988) Up- and down-modulation of single K^+ channel function by distinct second messengers. Trends Neurosci. 11:232–238

De Peyer, J.E., Cachelin, A.B., Levitan, I.B. & Reuter, H. (1982) Ca^{2+}-activated K^+ conductance in internally perfused snail neurons is enhanced by protein phosphorylation. Proc. Natl. Acad. Sci. USA 79:4207–4211

Fesenko, E.E., Kolesnikov, S.S. & Lynbarky, A.L. (1985) Induction by cyclic GMP of cationic conductance in plasma membrane of retinal rod outer segment. Nature (London) 313:310–313

Hamill, O.P., Marty, A., Neher, E., Sakmann, B. & Sigworth, F.J. (1981) Improved patch-clamp techniques for high-resolution current recording from cells and cell-free membrane patches. Pflüger's Arch. 391:85–100

Kaczmarek, L.K. (1987) The role of protein kinase C in the regulation of ion channels and neurotransmitter release. Trends Neurosci. 10:30–34
Kameyama, M., Hescheler, J., Hofmann F. & Trautwein, W. (1986) Modulation of Ca current during the phosphorylation cycle in the guinea pig heart. Pflüger's Arch. 407:123–128
Kandel, E.R. & Schwartz, J.H. (1982) Molecular biology of learning: modulation of transmitter release. Science 218:433–443
Levitan, I.B. (1985) Phosphorylation of ion channels. J. Membrane Biol. 87:177–190
Levitan, I.B. (1988) Modulation of ion channels in neurons and other cells. Annu. Rev. Neurosci. 11:119–136
Neer, E.J. & Clapham, D.E. (1988) Roles of G protein subunits in transmembrane signalling. Nature (London) 333:129–134
Noma, A. (1983) ATP-regulated K^+ channels in cardiac muscle. Nature (London) 305:147–148
Nowycky, M.C., Fox, A.P. & Tsien, R.W. (1985) Three types of neuronal calcium channel with different calcium agonist sensitivity. Nature (London) 316:440–443
Reuter, H. (1983) Calcium channel modulation by neurotransmitters, enzymes and drugs. Nature (London) 301:569–574
Reuter, H. & Porzig, H. (1988) Calcium Channels: Diversity and Complexity. Nature (London) 336:113–114
Reuter, H., Kokubun, S. & Prod'hom, B. (1986) Properties and modulation of cardiac calcium channels. J. Exp. Biol. 124:191–201
Rosenthal, W., Hescheler, J., Trautwein, W. & Schultz G. (1988) Control of voltage-dependent Ca^{2+} channels by G protein-coupled receptors. FASEB J. 2:2784–2790
Sigel, E. & Baur, R. (1988) Activation of protein kinase C differentially modulates neuronal Na^+, Ca^{2+}, and γ-aminobutyrate type A channels. Proc. Natl. Acad. Sci. USA 85:6192–6196
Tsien, R.W., Bean, B.P., Hess, P., Lansmann, J.B., Nilius, B. & Nowycky, M.C. (1986) Mechanisms of calcium channel modulation by β-adrenergic agents and dihydropyridine Ca agonists. J. Cell. Mol. Cardiol. 18:691–710

The Structure of the Skeletal and Nonskeletal Muscle Calcium Channel

F. HOFMANN[1], P. RUTH[1], T. SCHNEIDER[1], J. FELBEL[1], S. REGULLA[1], M. BIEL
W. NASTAINCZYK[1], and V. FLOCKERZI[1]

1 Introduction

A number of fundamental biological processes such as mobility, secretion, and neurotransmission are regulated by a change in the cytosolic calcium concentration. The cytosolic calcium concentration is raised from around 0.1 to 1 μM either by a release from cellular stores or by an influx from the extracellular space through voltage-dependent calcium channels. These channels are located in the plasma and transverse tubular membranes of many cells and are classified into at least three types (Nowycky et al. 1985), the T- (*t*iny or *t*ransient), N- (*n*either T- *n*or L-, or *n*euronal), and L- (*l*arge or *l*ong-*l*asting) channels. The L-type channels are the target of different organic drugs, the calcium channel blockers, which are used therapeutically in a variety of cardiovascular diseases. L- and N-type channels are not only regulated by the membrane potential or drugs but also by hormones through phosphorylation and/or G proteins. The electrophysiological properties of the L-type channel have been studied in great detail in cardiac, neuronal, and smooth muscle cells. In contrast, its biochemical and molecular structure has been elucidated up to now only in skeletal muscle.

2 Structure of the Skeletal Muscle Calcium Channel

The transversal tubular membranes of the skeletal muscle contain high affinity binding sites for dihydropyridines and phenylalkylamines (Galizzi et al. 1986). The high concentration of the transversal tubular binding sites has facilitated their purification to apparent homogeneity (Curtis and Catterall 1984; Flockerzi et al. 1986a; Takahashi et al. 1987; Leung et al. 1988). The purified channel contains three proteins of 165 kDa (α_1), 55 kDa (β), and 32 kDa (γ) which copurify in a constant 1:1:1 ratio (Sieber et al. 1987). A further glycoprotein which contains two disulfide-linked peptides of 135 kDa (α_2) and 28 kDa (δ) is present in the purified preparation. The relationship of this protein to the other three subunits is unknown.

The oligomeric structure of the channel is unclear at present. Electron microscopy of the purified channel proteins suggests that at least the α_1-, β-, and γ-subunits are part of a higher ordered structure (Leung et al. 1988). The α_1-subunit contains the binding sites for calcium channel blockers (Sieber et al. 1987; Striessnig et al. 1986a, 1987;

[1]Institut für Physiologische Chemie der Universität des Saarlandes, Medizinische Fakultät, D-6650 Homburg-Saar, Fed. Rep. of Germany

40. Colloquium Mosbach 1989
Molecular Mechanisms of Hormone Action

Takahashi et al. 1987), is readily phosphorylated by cAMP-kinase at Ser 687 (Röhrkasten et al. 1988), and has the primary structure of a voltage-dependent ion channel (Tanabe et al. 1987; Ellis et al. 1988). Antibodies specific for the α_1-, β-, and γ-subunit modulate the calcium channel of cultured cells and the reconstituted skeletal muscle channel supporting an oligomeric structure of the channel and a modulatory role for the β- and γ-subunit (Campbell et al. 1988; Morton et al. 1988; Vilven et al. 1988).

The primary sequence of the α_1-, α_2-, and β-subunit has been elucidated recently by cloning of the corresponding cDNAs. The structure of the γ-subunit is not known at present. The α_1-subunit contains four transmembrane regions which are composed of six transmembrane α-helices (Tanabe et al. 1987). The same topology is found in other voltage-dependent ion channels. The deduced amino acid sequence of the α_2 protein is compatible with that of a peripheral membrane protein, although it contains three transmembrane α-helices (Ellis et al. 1988). In agreement with its behavior as glycoprotein several consensus sequences for glycosylation are present in a large extracellular domain of the protein. The β-subunit contains four hydrophilic, 27 to 42 amino acid long α-helices, each of which contains a homologous stretch of eight amino acids (Ruth et al. 1989). The deduced amino acid of this protein is compatible with that of a peripheral membrane protein. The deduced amino acid sequence contains several potential phosphorylation sites. One of these sites, Ser 182, is phosphorylated in vitro by cAMP-kinase. A second and third site may be phosphorylated by cGMP-kinase and protein kinase C which enzymes modify rapidly the β-subunit in vitro (Nastainczyk et al. 1987; Jahn et al. 1988).

3 Binding Characteristics of the Purified Channel

Only the 165 kDa (α_1) subunit binds the dihydropyridines, phenylalkylamines, and benzothiazepines with high affinity (Flockerzi et al. 1986a; Striessnig et al. 1986b; Sieber et al. 1987). The respective apparent binding constants of the purified receptor are two- to tenfold higher than those of the membrane-bound receptor. The decreased affinity is due mainly to the difficulty in separating rapidly bound and free ligand when the solubilized receptor is used. The binding of dihydropyridines and phenylalkylamines is modulated by calcium. The high affinity binding of PN 200–110, a dihydropyridine, occurs only in the presence of micromolar to millimolar concentrations of calcium. The binding of PN 200–110 is regulated allosterically further by diltiazem and verapamil. In contrast, binding of devapamil, a verapamil analog, occurs in the absence of calcium if carried out at 4°C. Micromolar concentrations of calcium inhibit the binding of devapamil at 4°C. However, if the binding reaction is carried out at 30°C, devapamil binds only in the presence of a micromolar concentration of calcium and binding is inhibited by millimolar calcium. Preliminary analysis suggests that 0.1 to 1.0 μM calcium decreases the dissociation rate of devapamil at 30°C. Preincubation of the receptor for less than 20 min at 30°C in the absence of calcium and devapamil prevents binding of devapamil at 4°C. Apparently, a low concentration of calcium is necessary to stabilize the receptor at 30°C in a conformation which allows high affinity binding of devapamil. This apparent temperature dependence is reminiscent of the in vivo situation where the charge transfer of skeletal muscle is blocked by calcium channel blockers only at low temperatures and/or after partial depolarization of the plasma membrane (Berwe et al. 1987; Lamb and Walsh 1987; Rios and Brum 1987). Inactivation of the L-type calcium channels, which is induced by an increase in cytosolic calcium, increases the affinity of calcium channel blockers by several orders of magnitude (Bean 1984). This suggests that the conformation of the purified receptor protein complex may fluctuate between two states which are analogous to an active and inactivated channel.

4 Reconstitution of a Calcium Channel

The T-tubular localization of the calcium channel blocker binding sites has prevented so far a clear identification of their function. These binding sites may function only as a charge carrier which links membrane depolarization with the intracellular calcium release. Myotubes and single cells contain calcium channels which are similar to L-type channels of cardiac myocytes. T-tubular membranes contain L-type channels if the membrane is reconstituted into a phospholipid bilayer. The open probability of these channels increases after the addition of the calcium channel agonist Bay K 8644, cAMP-kinase and ATP, and the α-subunit of the GTP-binding protein of G_s (Flockerzi et al. 1986b; Talvenheimo et al. 1987; Hymel et al. 1988; Ma and Coronado 1988; Pelzer et al. 1988; Yatani et al. 1988). The channels are blocked by organic and inorganic calcium channel blockers. Similarly, the purified receptor complex reconstitutes to an L-type calcium channel. The opening probability increases after the addition of the channel agonist Bay K 8644, cAMP-kinase, and ATP•Mg. The channel is blocked by organic and inorganic calcium channel blockers and has a single channel conductance of 20 pSi (Flockerzi et al. 1986b; Pelzer et al. 1988). However, the reconstituted proteins support also channels of lower conductance which are not regulated by phosphorylation (Pelzer et al. 1988). The channel-forming protein is the α_1-subunit since (1) reconstitution of an isolated α_1-subunit leads to the formation of an L-type calcium channel (Pelzer et al. 1988); (2) injection of the cDNA for the α_1-subunit into dysgenic muscle cells, which lack the α_1-subunit, restores E-C coupling and a dihydropyridine-sensitive slow calcium channel in these cells (Tanabe et al. 1988); (3) antisense RNA complementary to the cDNA of the skeletal muscle α_1-subunit cDNA prevents the expression of cardiac muscle L-type channel in oocytes (Lotan et al. 1989). Thus, it is quite clear that the α_1-subunit is a calcium channel. This channel has slower opening and closing kinetics than the channel from other tissues, suggesting that the skeletal muscle channel differs from that of other tissues.

5 Structure of Nonskeletal Muscle L-Type Calcium Channels

The identity of the L-type calcium channel in other tissues has not been elucidated completely. Photoaffinity labeling of purified protein fractions and membranes suggests that the cardiac muscle receptor for calcium channel blockers is slightly larger than that of skeletal muscle. Apparent molecular weights of 195 versus 165 kDa have been reported (Schneider and Hofmann 1988; Chang and Hosëy 1988). A similar molecular weight has been observed for the brain receptor (Striessnig et al. 1988).

Northern blots carried out with an oligonucleotide probe derived from the cDNA of the skeletal muscle α_1-subunit show cross-hybridization to an 8.2 kb mRNA in brain, heart, and smooth muscle and to a 6.3 kb mRNA in skeletal muscle, suggesting again that the skeletal muscle channel differs from that of other excitable tissues. Monoclonal antibodies against the α_1-subunit of the rabbit skeletal muscle channel do not specifically label proteins of other tissues. These antibodies detect the α_1-subunit in the skeletal muscle from human, rabbit, guinea pig, hamster, cow, pig, chick, and frog. In contrast, antibodies directed against the α_2-proteins of the rabbit skeletal muscle receptor label an identical or similar protein in a wide variety of tissues, including skeletal muscle, heart, brain, and smooth muscle (Schmid et al. 1986; Norman et al. 1987). Similarly, the cDNA of the α_2-protein hybridizes with the mRNA of other tissues in Northern blots (Ellis et al. 1988). This supports again the notion that an identical α_2-protein is present in all tissues, whereas the α_1-subunit differs quite considerably from that of skeletal muscle. So far, it is unclear whether the β- and γ-subunit are part of the calcium channel of other tissues. Preliminary data suggest that mRNA and a protein similar to the β-subunit of skeletal muscle may be present in brain. It is therefore possible that both subunits are present in other tissues. The proteins associated with the N-type channel are distinct from these proteins. N-type channels bind specifically ω-conotoxin and have a reported molecular weight of above 200 kDa (Abe and Saisu 1987; Barhanin et al. 1988; Marqueze et al. 1988).

6 In Vivo Regulation and Conclusions

Calcium channels are regulated in vivo by three mechanisms: The membrane potential, protein kinases, and G proteins. The latter two mechanisms are of interest since they allow hormonal regulation of channel opening and closing. Cyclic AMP-dependent protein kinase increases the open probability of the cardiac and some neuronal L-type channels in vivo (Trautwein et al. 1986; Gray and Johnston 1987). The protein phosphorylated in vivo is unknown but the work with the skeletal muscle channel suggests that it may be the α_1-subunit. The α-subunits of G_o, G_i, and G_s appear to regulate the opening and closing probability of calcium channels in neuronal, neurosecretory, and cardiac cells (Yatani et al. 1987; Rosenthal et al. 1988). The exact mechanism has not been determined in most cases. The opening probability of reconstituted cardiac and skeletal muscle calcium channels increases after addition of the α-subunit of G_s. However, it is unclear whether the α_1-subunit of G_s directly affects a cardiac calcium channel in vivo, since inhibition of cAMP-kinase blocks the β-adrenergic stimulation of the cardiac calcium current. However, recently it has been found that the β-adrenergic receptor of isolated tracheal smooth muscle cells stimulates the calcium current (Felbel et al. 1989). This stimulation was not mimicked by the catalytic subunit of cAMP-kinase but was mimicked by GTPγS. This finding suggests that in these smooth muscle cells the α-subunit of G_s may couple directly to the calcium channel. These results support therefore the hypothesis that L-type calcium channels are regulated by multiple mechanisms in vivo.

Acknowledgments. We thank Mrs. Sommer for typing the manuscript. The experimental work of the authors was supported by grants of the Fonds der chemischen Industrie and Deutsche Forschungsgemeinschaft.

References

Abe, T. & Saisu, H. (1987) Identification of the receptor for ω-conotoxin in brain. J. Biol. Chem. 262:9877–9882

Barhanin, J., Schmid, A. & Lazdunski, M. (1988) Properties of structure and interaction of the receptor for ω-conotoxin, a polypeptide active on Ca^{2+} channels. Biochem. Biophys. Res. Commun. 150: 1051–1062

Bean, B.P. (1984) Nitrendipine block of cardiac calcium channels: high-affinity binding to the inactivated state. Proc. Natl. Acad. Sci. USA 81:6388–6392

Berwe, D., Gottschalk, G. & Lüttgau, C.H. (1987) Effects of the calcium antagonist gallopamil (D600) upon excitation-contraction coupling in toe muscle fibres of the frog. J. Physiol. 385:693–707

Campbell, K.P., Leung, A.T., Sharp, A.H., Imagawa, T. & Kahl, S.D. (1988) Ca^{2+} channel antibodies: subunit-specific antibodies as probes for structure and function. In: Morad M., Nayler, W.G., Kazda, S. & Schramm, M. (eds.) The calcium channel: structure, function and implication. Springer, Berlin Heidelberg New York Tokyo, pp. 586–600

Chang, C.F. & Hosey, M.M. (1988) Dihydropyridine and phenylalkylamine receptors associated with cardiac and skeletal muscle calcium channels are structurally different. J. Biol. Chem. 263: 18929–18937

Curtis, B.M. & Catterall, W.A. (1984) Purification of the calcium antagonist receptor of the voltage-sensitive calcium channel from skeletal muscle transverse tubules. Biochemistry 23:2113–2118

Ellis, S.B., Williams, M.E., Ways, N.R., Brenner, R., Sharp, A.H., Leung, A.T., Campbell, K.P., McKenna, E., Koch, W.J., Hui, A., Schwartz, A. & Harpold, M.M. (1988) Sequence and expression of mRNAs encoding the α_1 and α_2 subunits of a DHP-sensitive calcium channel. Science 241:1661–1664

Felbel, J., Welling, A., Peper, K. & Hofmann, F. (1989) Activation of the β-adrenergic receptor increases L-type calcium current independent of cAMP and cAMP-kinase in adult tracheal smooth mucle cells. Biol. Chem. Hoppe-Seyler 370:787

Flockerzi, V., Oeken, H.-J. & Hofmann, F. (1986a) Purification of a functional receptor for calcium channel blockers from rabbit skeletal muscle microsomes. Eur. J. Biochem. 161:217–224

Flockerzi, V., Oeken, H.-J., Hofmann, F., Pelzer, D., Cavalié, A. & Trautwein, W. (1986b) The purified dihydropyridine binding site from skeletal muscle T-tubules is a functional calcium channel. Nature (London) 323:66–68

Galizzi, D.-P., Borsotto, M., Barhanin, J., Fosset, M. & Lazdunski, M. (1986) Characterization and photoaffinity labeling of receptor sites for the Ca^{2+} channel inhibitors d-cis-diltiazem, (±)-bepridil, desmethoxyverapamil, and (+)-PN 200–110 in skeletal muscle transverse tubule membranes. J. Biol. Chem. 261:1393–1397

Gray, R. & Johnston, D. (1987) Noradrenaline and beta-adrenoceptor agonists increase activity of voltage-dependent calcium channels in hippocampal neurons. Nature (London) 327:620–622

Hymel, L., Striessnig, J., Glossmann, H. & Schindler, H. (1988) Purified skeletal muscle 1,4-dihydropyridine receptor forms phosphorylation-dependent oligomeric calcium channels in planar bilayers. Proc. Natl. Acad. Sci. USA 85:4290–4294

Jahn, H., Nastainczyk, W., Röhrkasten, A., Schneider, T. & Hofmann, F. (1988) Site specific phosphorylation of the purified receptor for calcium channel blockers by cAMP-, cGMP-dependent protein kinase, protein kinase C, calmodulin-dependent protein kinase II, and casein kinase II. Eur. J. Biochem. 178:535–542

Lamb, G.D. & Walsh, T. (1987) Calcium currents, charge movement and dihydropyridine binding in fast- and slow-twitch muscle of rat and rabbit. J. Physiol. 393:595–617

Leung, A.T., Imagawa, T., Block, B., Franzini-Armstrong, C. & Campbell, K.P. (1988) Biochemical and ultrastructural characterization of the 1,4-dihydropyridine receptor from rabbit skeletal muscle. J. Biol. Chem. 263:994–1001

Lotan, I., Goelet, P., Gigi, A. & Dascal, N. (1989) Specific block of calcium channel expression by a fragment of dihydropyridine receptor cDNA. Science 243:666–669

Ma, J. & Coronado, R. (1988) Heterogeneity of conductance states in calcium channels of skeletal muscle. Biophys. J. 53:387–395

Marqueze, B., Martin-Moutot, N., Leveque, C. & Couraud, F. (1988) Characterization of the ω-conotoxin-binding molecule in rat brain synaptosomes and cultured neurons. Mol. Pharmacol. 34:87–90

Morton, M.E., Caffrey, J.M., Brown, A.M. & Froehner, S.C. (1988) Monoclonal antibody to the α1 subunit of the dihydropyridine-binding complex inhibits calcium currents in BC3H1 myocytes. J. Biol. Chem. 263:613–616

Nastainczyk, W., Röhrkasten, A., Sieber, M., Rudolph, C., Schächtele, C., Marme, D. & Hofmann, F. (1987) Phosphorylation of the purified receptor for calcium channel blockers by cAMP kinase and protein kinase C. Eur. J. Biochem. 169:137–142

Norman, R.I., Burgess, A.J., Allen, E. & Harrison, T.M. (1987) Monoclonal antibodies against the 1,4-dihydropyridine receptor associated with voltage-sensitive Ca^{2+} channels detect similar polypeptides from a variety of tissues and species. FEBS Lett. 212:127–132

Nowycky, M.C., Fox, A.P. & Tsien, R.W. (1985) Three types of neuronal calcium channel with different calcium agonist sensitivity. Nature (London) 316:440–443

Pelzer, D., Cavalié, A., Flockerzi, V., Hofmann, F. & Trautwein, W. (1988) Reconstitution of solubilized and purified dihydropyridine receptor from skeletal muscle microsomes as two single calcium channel conductances with different functional properties. In: Morad, M., Nayler, W.G., Kazda, S., Schramm, M.(eds.) The calcium channel: structure, function and implication. Springer, Berlin Heidelberg New York Tokyo, pp 217–230

Rios, E. & Brum, G. (1987). Involvement of dihydropyridine receptors in excitation-contraction coupling in skeletal muscle Nature (London) 235:717–720

Röhrkasten, A., Meyer, H., Nastainczyk, W., Sieber, M. & Hofmann, F. (1988) cAMP-dependent protein kinase rapidly phosphorylates Ser 687 of the rabbit skeletal muscle receptor for calcium channel blockers. J. Biol. Chem. 263:15325–15329

Rosenthal, W., Hescheler, J., Trautwein, W. and Schultz, G. (1988) Control of voltage-dependent Ca^{2+} channels by G protein-coupled receptors. FASEB J. 2:2784–2790

Ruth, P., Röhrkasten, A., Biel, M., Bosse, E., Regulla, S., Meyer, H.E., Flockerzi, V. & Hofmann, F. (1989) Primary structure of the β-subunit of the dhp-sensitive calcium channel from skeletal muscle. Science, in press

Schmid, A., Barhanin, J., Coppola, T., Borsotto, M. & Lazdunski, M. (1986) Immunochemical analysis of subunit structures of 1,4-dihydropyridine receptor associated with voltage-dependent Ca^{2+} channels in skeletal, cardiac, and smooth muscles. Biochemistry 25:3492–3495

Schneider, T. & Hofmann, F. (1988) The bovine cardiac receptor for calcium channel blockers is a 195 kDa protein. Eur. J. Biochem. 174:127–135

Sieber, M., Nastainczyk, W., Zubor, V., Wernet, W. & Hofmann, F. (1987) The 165-kDa peptide of the purified skeletal muscle dihydropyridine receptor contains the known regulatory sites of the calcium channel. Eur. J. Biochem. 167:117–122

Striessnig, J., Moosburger, K., Goll, D., Ferry, D.R. & Glossmann, H. (1986a) Stereoselective photoaffinity labelling of the purified 1,4-dihydropyridine receptor of the voltage-dependent calcium channel. Eur. J. Biochem. 161:603–609

Striessnig, J., Goll, A., Moosburger, K. & Glossmann, H. (1986b). Purified calcium channels have three allosterically coupled drug receptors. FEBS Lett. 197:204–210

Striessnig, J., Knaus, H.-G., Grabner, M., Moosburger, K., Seitz, W., Lietz, H. & Glossman, H. (1987) Photoaffinity labelling of the phenylalkylamine receptor of the skeletal muscle transverse-tubule calcium channel. FEBS Lett. 212:247–253

Striessnig, J., Knaus, H.-G. & Glossmann, H. (1988) Photoaffinity-labelling of the calcium-channel-associated 1,4-dihydropyridine and phenylalkylamine receptor in guinea-pig hippocampus. Biochem. J. 253:39–47

Takahashi, M., Seagar, M.J., Jones, J.F., Reber, B.F.X. & Catterall, W.A. (1987) Subunit structure of dihydropyridine-sensitive calcium channel from skeletal muscle. Proc. Natl. Acad. Sci. USA 84:5478–5482

Talvenheimo, J.A., Worley, III J.F. & Nelson M.T. (1987) Heterogeneity of calcium channels from a purified dihydropyridine receptor preparation. Biophys. J. 52:891–899

Tanabe, T., Takeshima, H., Mikami, A., Flockerzi, V., Takahashi, H., Kangawa, K., Kojima, M., Matsuo, H., Hirose, T. & Numa, S. (1987) Primary structure of the receptor for calcium channel blockers from skeletal muscle. Nature (London) 328:313–318

Tanabe, T., Beam, K.G., Powell, J.A. & Numa, S. (1988) Restoration of excitation-contraction coupling and slow calcium current in dysgenic muscle by dihydropyridine receptor complementary DNA. Nature (London) 366:134–139

Trautwein, W., Kameyama, M., Hescheler, J. & Hofmann, F. (1986) Cardiac calcium channels and their transmitter modulation. Progr. Zool. 33:163–182

Vilven, J., Leung, A.T., Imagawa, T., Sharp, A.H., Campbell, K.P. & Coronado, R. (1988) Interaction of calcium channels of skeletal muscle with monoclonal antibodies specific for its dihydropyridine receptor. Biophys. J. 53:556a

Yatani, A., Codina, J., Imoto, Y., Reeves, J.P., Birnbaumer, L. & Brown, A.M. (1987) A G-protein directly regulates mammalian cardiac calcium channels. Science 238:1288–1292

Yatani, A., Imoto, Y., Codina, J., Hamilton, S.L., Brown, A.M. & Birnbaumer, L. (1988). The stimulatory G-protein of adenylyl cyclase, Gs, also stimulates dihydropyridine-sensitive Ca^{2+} channels. J. Biol. Chem. 263:9887–9895

Involvement of Pertussis Toxin-Sensitive G-Proteins in the Modulation of Voltage-Dependent Ca^{2+} Channels by Extracellular Signals

W. ROSENTHAL[1], S. OFFERMANNS[1], J. HESCHELER[1], K. SPICHER[1], K.-D. HINSCH[1], U. RUDOLPH[1], and G. SCHULTZ[1]

1 Introduction

Heterotrimeric guanine nucleotide-binding proteins (G-proteins; see Chap. 9, this Vol.) are located at the inner face of the plasma membrane. They act as transducer molecules by coupling transmembranous receptors to effectors which generate intracellular signals. Well-studied effectors are two enzymes, the ubiquitous adenylate cyclase and the retinal phosphodiesterase (see Chap. 12, this Vol.). In both instances, the G-protein α-subunit, activated by a transmembranous receptor, changes the enzyme activity by direct interaction with the effector molecule.

G-proteins are also involved in the modulation of ion channels. There are many examples for a *distant control of ion channels by G-proteins.* Here, a signal generated by an enzyme either affects ion channels directly, for example, in rod outer segments (Hanke et al. 1988), or the signal stimulates a protein kinase which, in turn, changes the activity of ion channels by phosphorylation; well-studied examples are cardiac Ca^{2+} channels (see Chap. 16, this Vol.).

So far, a direct interaction of G-protein subunits with ion channels has not been demonstrated. There is, however, increasing evidence for a *close, membrane-confined control of ion channels by G-proteins.* In this contribution, we summarize data we have obtained with three different cell types. These data support the hypothesis that pertussis toxin-sensitive G-proteins exert both a close stimulatory and a close inhibitory control of voltage-dependent Ca^{2+} channels (for recent reviews, see Rosenthal et al. 1988b; Dolphin 1990; Schultz et al. 1990). The close control of cardiac voltage-dependent Ca^{2+} channels by the cholera toxin-sensitive G-protein, G_s, and the close control of K^+ channels by G-proteins in cardiac, endocrine, and neuronal cells are subject of the contribution by Birnbaumer (Chap. 18, this Vol.).

2 Methods

Electrophysiological experiments reported here were performed in the whole-cell configuration of the patch-clamp technique (Hamill et al. 1981). Patch pipettes had a tip opening of 1 to 3 μm and a resistance of 2–5 MΩ. Ca^{2+} inward currents were measured under voltage-clamp conditions in the presence of Cs^+, using Ba^{2+} (10.8

[1]Institut für Pharmakologie, Freie Universität Berlin, Thielallee 69–73, D-1000 Berlin 33, Fed. Rep. of Germany

40. Colloquium Mosbach 1989
Molecular Mechanisms of Hormone Action

mM) as charge carrier. Ca^{2+} inward currents were evoked by depolarizing potential steps from −40 to 0 mM (for details of electrophysiological and biochemical methods, see Hescheler et al. 1987, 1988a,b; Rosenthal et al. 1988a; Offermanns et al. 1989).

3 Results and Discussion

3.1 Inhibitory Control of Ca^{2+} Currents in Nueroblastoma × Glioma Hybrid Cells

Elevation of cytosolic Ca^{2+}, required for secretion of neurotransmitters, is accomplished by depolarization of the plasma membrane and subsequent opening of voltage-dependent Ca^{2+} channels. There are at least two mechanisms by which agonists acting on inhibitory presynaptic receptors reduce Ca^{2+} influx through voltage-dependent Ca^{2+} channels and, thereby, inhibit secretion of nuerotransmitters. (1) Agonists can directly inhibit voltage-dependent Ca^{2+} channels, i.e., they reduce the flow of Ca^{2+} through Ca^{2+} channels during a depolarizing pulse. (2) Agonists can stimulate neuronal K^+ channels, probably via the G-protein G_o (vanDongen et al. 1988). This leads to hyperpolarization of the plasma membrane and, as a consequence, to closure of voltage-dependent Ca^{2+} channels. Inhibitory agonists also reduce cytosolic cyclic AMP by inhibition of adenylate cyclase. Whether this effect, which is mediated by a G-protein of the G_i family, contributes to the inhibition of Ca^{2+} influx in neuronal cells, is not known.

To examine the effects of an inhibitory agonist on voltage-dependent Ca^{2+} currents, we chose as a model system the neuroblastoma × glioma (N × G) hybrid cell (108CC15; Hamprecht et al. 1985) which possesses a high number of opioid receptors of the δ-type. Correspondingly, we used an opioid which preferentially activates δ-type receptors, the synthetic peptide, D-alanine-D-leucine-enkephalin (DADLE). Addition of the opioid to the bath solution caused a rapid decrease of the Ca^{2+} current (Hescheler et al. 1987, 1988b). After removal of DADLE, the current immediately increased and reached control values within 1 min. Thus, the effect of DADLE was rapid in onset and fully reversible. The effect of the agonist was half-maximal at a concentration of 10 nM. The maximal inhibition obtained with DADLE was about 70%.

Having established the inhibitory effects of DADLE on Ca^{2+} currents, we performed experiments to obtain information on the underlying molecular mechanism (Table 1). Since DADLE inhibits adenylate cyclase in N × G cells, one possible mechanism is a decrease in cyclic AMP levels, leading to a reduced activity of cAMP-dependent protein kinase. However, intracellularly applied cyclic AMP or forskolin added to the bath solution affected neither Ca^{2+} currents in N × G cells nor interfered with the DADLE-induced inhibition of Ca^{2+} currents. Similarly, intracellularly applied cyclic GMP or extracellularly applied sodium nitroprusside were inactive. In accordance with the majority of reports, a protein kinase C-activating phorbol ester did not inhibit Ca^{2+} currents in N × G cells. It should be emphasized that a role of protein kinase C in the inhibition of neuronal Ca^{2+} currents by extracellular signals has been proposed by other investigators (Rane and Dunlap 1986; Rane et al. 1987; for recent reviews on this issue, see Rosenthal et al. 1988b; Dolphin 1990; Schultz et al. 1990). We have not tested the effects of arachidonic acid on Ca^{2+} currents in N × G

Table 1. Effects of intracellular signal molecules on receptor-modulated, voltage-dependent Ca^{2+} currents in neuroblastoma × glioma (N×G) hybrid cells and other neuronal cells

Agent	Effect on Ca^{2+} currents in N×G cells	Effect on Ca^{2+} currents in Other neuronal cells	
Cyclic AMP, 3-isobutyl-1-methylxanthin, forskolin	None	None	(Wanke et al. 1987; Dolphin et al. 1989; Kasai and Aosaki 1989)
Cyclic GMP, nitroprusside	None	None	(Kasai and Aosaki 1989)
Arachidonate, inhibitors of phospholipases	?	None	(Kasai and Aosaki 1989; Dolphin et al. 1989)
Activators/inhibitors of protein kinase C	None	None	(Wanke et al. 1987; Dolphin et al. 1989; Kasai and Aosaki 1989)
		Inhibition	(Rane and Dunlap 1986; Rane et al. 1987)

cells; this fatty acid may be liberated by receptor-mediated activation of a phospholipase A_2. However, according to two recent reports (Kasai and Aosaki 1989; Dolphin et al. 1989), arachidonic acid or inhibitors of phospholipases do not modulate Ca^{2+} currents in neuronal cells. These results indicate that intracellular signal molecules or protein kinases activated by intracellular signal molecules are not involved in the inhibitory modulation of Ca^{2+} currents in N × G cells or in neuronal cells.

Effects of guanine nucleotides on N × G cell Ca^{2+} currents provided strong evidence for the involvement of a G-protein in Ca^{2+} current modulation. Intracellular application of the GTP analog, guanosine-5′-0-(3-thiotriphosphate) (GTPγS), which activates G-proteins in the absence of receptor agonists, caused a slow inhibition of Ca^{2+} currents to an extent similar to that induced by DADLE, if applied at a high concentration. At a low concentration (1 μM), GTPγS itself was without effect; if, however, DADLE was added to the bath solution, a strong inhibition of Ca^{2+} currents was observed, which persisted after removal of the agonist from the bath solution. Thus, DADLE, promoted the irreversible inhibition induced by GTPγS. In contrast, intracellular application of the GDP analog, guanosine-5′-0-(2-thiodiphosphate) (GDPβS), prevented the DADLE-induced inhibition of Ca^{2+} currents. The effects of guanine nucleotides on Ca^{2+} currents closely resemble their effects on the G protein-regulated enzyme, adenylate cyclase, and suggest the involvement of a G-protein.

The α-subunits of some G-proteins, namely, G_i and G_o, are substrates for the main exotoxin of *Bordetella pertussis,* pertussis toxin, which catalyzes ADP-ribosylation of G-protein α-subunits and, thereby, prevents activation of G-proteins by agonist-occupied receptors (Ui et al. 1984). After treatment of N × G cells with pertussis toxin (100 ng ml^{-1}) fo several hours, Ca^{2+} currents were not affected by extracellularly applied DADLE but remained fully responsive to intracellularly applied GTPγS. The results show that a pertussis toxin-sensitive G-protein is involved in Ca^{2+} current inhibition. In addition, they confirm the observation made in other systems (e.g., the adenylate cyclase system; Jakobs et al. 1984) that pertussis toxin prevents the receptor-mediated activation of G-proteins; it, however, does not prevent the receptor-independent activation of G-proteins by a GTP analog such as GTPγS.

Like membranes from brain (Mumby et al. 1988), membranes of N×G cells possess at least three pertussis toxin-sensitive G-proteins, i.e., two G-proteins of the G_i family with α-subunits of 40 and 41 kDa and G_o with a 39 kDa α-subunit (own unpublished data; Mullaney et al. 1988; Rosenthal et al. 1988a). Whereas the subtype of the 41 kDa G_i α-subunit is not known, the 40 kDa α-subunit corresponds to the G_{i2} α-subunit (own unpublished data). To identify the G-protein(s) activated by δ-receptors in membranes of N×G cells, we used the photoreactive GTP analog, [α-^{32}P]GTP azido-anilide ([α-^{32}P]AA-GTP; Pfeuffer 1977; Schäfer et al. 1988). Activation of G-proteins by agonist-occupied receptors should promote the release of GDP and allow the GTP analog to bind to the α-subunit. Experiments performed with purified G-proteins showed that, subsequent to radiation with ultraviolet light, the GTP analog was covalently attached to G-protein α-subunits but not to other G-protein subunits and that the reaction was specific for guanine nucleotides. Under control conditions, [α-^{32}P]AA-GTP was incorporated in membrane proteins of N×G cells, comigrating with the α-subunits of G_{i2} (40 kDa) and G_o (39 kDa). DADLE effectively stimulated photolabeling of both proteins. Somatostatin, a peptide hormone, known to inhibit Ca^{2+} currents in N×G cells (Tsunoo et al. 1986; own unpublished data) and pituitary cells (Lewis et al. 1986; Rosenthal et al. 1988a), had a qualitatively similar effect but was less potent than was DADLE. In contrast, bradykinin, which does not affect Ca^{2+} currents in N×G cells (unpublished), mainly affected photolabeling of the 40 kDa protein. Similarly, adrenaline, a weak inhibitor of Ca^{2+} currents in N×G cells (Brown et al. 1989; own unpublished data), preferentially stimulated photolabeling of the 40 kDa protein. We conclude from these experiments that receptors mediating a marked inhibition of Ca^{2+} currents are effective activators of endogenous G_o.

We also performed reconstitution experiments to identify the G-protein involved in inhibitory Ca^{2+} current modulation. For this purpose, we treated N×G cells with pertussis toxin for several hours. After this treatment, purified G-proteins were infused into cells via the patch pipette and the effect of extracellularly applied DADLE on voltage-dependent Ca^{2+} currents was studied. Intracellular application of G_i (possibly representing mainly G_{i1}) purified from porcine brain restored the inhibitory DADLE effect partially, if employed at a concentration of 4 nM; it was virtually inactive at a concentration of 0.4 nM. In contrast, the α-subunit of G_o, identified by highly specific antisera raised against synthetic peptides, fully restored the DADLE-induced inhibition, if employed at a concentration of 4 nM with or without a tenfold excess of G-protein $\beta\gamma$-complexes; its reconstituting activity was also obvious at a concentration of 0.4 nM. Thus, the data obtained in reconstitution experiments correlate well with data obtained with the photoreactive GTP analog and suggest a role for G_o in the inhibition of voltage-dependent Ca^{2+} channels.

3.2 *Bidirectional Control of Ca^{2+} Currents in a Pituitary Cell Line (GH_3)*

Pertussis toxin- and guanine nucleotide-sensitive inhibition of voltage-dependent Ca^{2+} currents has also been observed in a murine cell line (AtT-20) derived from the anterior pituitary (Lewis et al. 1986). In a pituitary cell line from rat (GH_3), we observed a dual control of Ca^{2+} currents (Rosenthal et al. 1988a). A secretion-stimulating hormone, luteinizing hormone-releasing hormone (LHRH) stimulated, whereas a secretion-inhibiting hormone, somatostatin, inhibited voltage-dependent Ca^{2+} currents. The effects of both hormones were rapid in onset and fully reversible. Both stimulatory and inhibitory hormonal modulations were independent of cyclo AMP and abolished by treatment of cells with pertussis toxin, indicating that not only inhibitory but also stimulatory Ca^{2+} current modulations are mediated by pertussis toxin-sensitive G-proteins.

This assumption is supported by data obtained with membranes from GH_3 cells. High-affinity GTPase in membrane preparations represents the enzymatic activity of G-proteins. The stimulatory hormones, LHRH and thyrotropin-releasing hormone (TRH), and the inhibitory hormone, somatostatin, stimulated high-affinity GTPases in membranes of GH_3 cells in a concentration-dependent manner (Offermanns et al. 1989). In membranes prepared from pertussis toxin-treated cells, the basal GTPase activity was lower than in control cells. In addition, the stimulatory effect of LHRH and that of TRH was largely reduced, and the stimulatory effect of somatostatin was abolished. The results show that the receptors for both stimulatory and inhibitory hormones interact with pertussis toxin-sensitive G-proteins. Since stimulation of phosphoinositide hydrolysis in pituitary cells by LHRH and TRH is not sensitive to pertussis toxin (Martin et al. 1986; Naor et al. 1986), this pathway is apparently not involved in the modulation of Ca^{2+} currents in this cell type.

For identification of G-proteins, we probed membranes of GH_3 cells with pertussis toxin and specific antisera. Very similar to membranes of neuronal cells, we found G-proteins of the G_i-type with α-subunits of 41 and 40 kDa; the latter corresponded to the α-subunit of G_{i2} (unpublished). We also found that a 39 kDa α-subunit was highly abundant in GH_3 cell membranes; this α-subunit was identified as G_o α-subunit (Rosenthal et al. 1988a). Inhibitory agonists, including carbachol and somatostatin, efficiently stimulated incorporation of the photoreactive GTP analog, [α-^{32}P]AA-GTP, into both 40 and 39 kDa proteins, which apparently represent the α-subunits of G_{i2} and G_o, respectively. In contrast, the stimulatory hormones, LHRH and TRH, preferentially stimulated photolabeling of the 40 kDa protein, presumably representing the G_{i2} α-subunit. These findings strongly suggest that hormonal stimulation and inhibition of voltage-dependent Ca^{2+} currents are mediated by distinct pertussis toxin-sensitive G-proteins.

3.3 Stimulatory Control of Ca^{2+} Currents in an Adrenocortical Cell Line (Y1)

Angiotensin II, the major stimulator of aldosterone secretion, caused a rapid and reversible stimulation of Ca^{2+} currents in the adrenocortical cell line, Y1 (Hescheler et al. 1988a). A stimulatory effect of this hormone on Ca^{2+} currents has also been observed in bovine glomerulosa cells (Cohen et al. 1988). In Y1 cells, angiotensin II increased Ca^{2+} currents independently of intracellularly applied cyclic AMP and cyclic GMP and independently of an extracellularly applied protein kinase C-activating phorbol ester. The effect was blocked by pertussis toxin, whereas the toxin does not affect hormonal stimulation of phosphoinositide hydrolysis in this cell type (Kojima et al. 1986).

In membranes of Y1 cells we detected pertussis-toxin substrates of about 40 kDa. With the help of specific antisera, these substrates were identified as G_i α-subunits. We, however, did not detect G_o in membranes of Y1 cells (Hescheler et al. 1988a; Rosenthal et al. 1988a).

4 Conclusions

Our data are consistent with the hypothesis that G_o mediates the pertussis toxin-sensitive inhibition of voltage-dependent Ca^{2+} currents. The experimental evidence for this assumption is summarized in Table 2.

The identity of the G-protein involved in the pertussis-toxin-sensitive stimulation of voltage-dependent Ca^{2+} currents in endocrine cells is not clear. As we only know of two groups of nonretinal G-proteins which are substrates for pertussis toxin, the G_i family and G_o, one may suggest that a G_i-type G-protein is a likely candidate. This assumption is supported by our finding that Y1 cells in which angiotensin II stimulates voltage-dependent Ca^{2+} currents in a pertussis toxin-sensitive manner possess G_i-type G-proteins but lack G_o.

So far, there is no convincing evidence for the involvement of cytosolic signal molecules or of protein kinases activated by cytosolic signal molecules in the pertussis toxin-sensitive hormonal modulations of voltage-dependent Ca^{2+} channels. Whether the hormonal control of voltage-dependent Ca^{2+} channels in neuronal and endocrine

Table 2. Experimental evidence for a physiological role for G_o in the inhibition of voltage-dependent Ca^{2+} channels by hormones and neurotransmitters

Occurrence of G_o in cells that exhibit pertussis toxin-sensitive, agonist-induced inhibition of Ca^{2+} currents

- Central and peripheral neurons (Asano et al. 1988)
- Neuroblastoma × glioma hybrid cells (Mullaney et al. 1988)
- Anterior pituitary (Asano et al. 1988)
- Pituitary GH_3 cells (Rosenthal et al. 1988a)

Reconstitution of agonist-induced Ca^{2+} current inhibition in pertussis toxin-treated cells by intracellular application of purified G_o

- Neuroblastoma × glioma hybrid cells (Hescheler et al. 1987, 1988b)
- Dorsal root ganglion neurons (Ewald et al. 1988; Lux et al. 1989)
- Snail neurons (Harris-Warrick et al. 1988)

Attenuation of agonist-induced Ca^{2+} current inhibition by intracellular application of antibodies against the G_o α-subunit

- Snail neurons (Harris-Warrick et al. 1988)
- Neuroblastoma × glioma hybrid cells (Brown et al. 1989)

Correlation of G_o α-subunit expression with the ability of dopamine to inhibit prolactin secretion

- Rat pituitary tumor cells (Collu et al. 1988)

Stimulation by inhibitory agonists of guanine nucleotide binding to a 39 kDa membrane protein comigrating with the G_o α-subunit

- Neuroblastoma × glioma hybrid cells (unpublished)
- Pituitary GH_3 cells (unpublished)

cells is accomplished by just three signal transduction components including receptors, pertussis toxin-sensitive G-proteins and channel proteins, or whether additional regulatory components are involved, remains to be clarified.

Acknowledgments. We thank Monika Bigalke, Evelyn Bombien, and Inge Reinsch for excellent technical assistance and the Deutsche Forschungsgemeinschaft for financial support.

References

Asano, T., Semba, R., Kamiya, N., Ogasawara, N. & Kato, K. (1988) G_o, a GTP-binding protein: immunohistochemical localization in the rat. J. Neurochem. 50:1164–1169

Brown, D., McFadzean & Milligan, G. (1989) Antibodies to the GTP-binding protein G_o attenuate the inhibition of the calcium current by noradrenaline in mouse neuroblastoma × rat glioma (NG 108-15) hybrid cells. J. Physiol. 415:20P (suppl.)

Cohen, C., McCarthy, R., Barrett, P. & Rasmussen, H. (1988) Ca channels in adrenal glomerulosa cells: K^+ and angiotensin II increase T-type Ca channel current. Proc. Natl. Acad. Sci. USA 85:2412–2416

Collu, R., Bouvier, C., Lagacé, G., Unson, C., Milligan, G., Goldsmith, P. & Spiegel, A. (1988) Selective deficiency of guanine nucleotide-binding protein G_o in two dopamine-resistant pituitary tumors. Endocrinology 122:1176–1178

Dolphin, A. (1990) G protein modulation of calcium currents in neurons. Annu. Rev. Physiol. 52 (in press)

Dolphin, A., McGuirk, S.M. & Scott, R.H. (1989) An investigation into the mechanisms of inhibition of calcium channel currents in rat cultured sensory neurones by guanine nucleotide analogues and (-)-baclofen. Br. J. Pharmacol. 97:263–273

Ewald, D.A., Sternweis, P.C. & Miller, R.J. (1988) Guanine nucleotide-binding protein G_o-induced coupling of neuropeptide Y receptors to Ca^{2+} channels in sensory neurons. Proc. Natl. Acad. Sci. USA 85:3633–3637

Hamill, O.P., Marty, A., Neher, E. & Sakmann, B. (1981) Improved patch-clamp techniques for high-resolution current recording from cells and cell-free membrane patches. Pflüger's Arch. 391:85–100

Hamprecht, B., Glaser, T., Reiser, G., Bayer, E. & Propst, F. (1985) Culture and characteristics of neuroblastoma × glioma hybrid cells. Meth. Emzymol. 109:316–341

Hanke, W., Cook, N.J. & Kaupp, U.B. (1988) cGMP-dependent channel protein from photoreceptor membranes: single-channel activity of the purified reconstituted protein. Proc. Natl. Acad. Sci. USA 85:94–98

Harris-Warrick, R., Hammond, C., Pauperdin-Trisch, D., Homburger, V., Rouot, B., Bockaert, J. & Gerschenfeld, H. (1988) An α_{40} subunit of a GTP-binding protein immunologically related to G_o mediates a dopamine-induced decrease of Ca^{2+} current in snail neurons. Neuron 1:27–32

Hescheler, J., Rosenthal, W., Trautwein, W. & Schultz, G. (1987) The GTP-binding protein, G_o, regulates neuronal calcium channels. Nature (London) 325:445–447

Hescheler, J., Rosenthal, W., Hinsch, K.-D., Wulfern, M., Trautwein, W. & Schultz, G. (1988a) Angiotensin II-induced stimulation of voltage-dependent calcium channels in an adrenal cortical cell line. EMBO J. 7:619–624

Hescheler, J., Rosenthal, W., Wulfern, M., Tang, M., Yajima, M., Trautwein, W. & Schultz, G. (1988b) Involvement of the guanine nucleotide-binding protein, N_o, in the inhibitory regulation of neuronal calcium currents. Adv. second Messenger Phosphoprotein Res. 21:165–174

Jakobs, K.H., Aktories, K. & Schultz, G. (1984) Mechanism of pertussis-toxin action on the adenylate cyclase system. Eur. J. Biochem. 140:177–181

Kasai, H. & Aosaki, T. (1989) Modulation of Ca-channels current by an adenosine analog mediated by a GTP-binding protein in chick sensory neurons. Pflüger's Arch. 414:145–149

Kojima, I., Shibata, H., Ogata, E. (1986) Pertussis toxin blocks angiotensin II-induced calcium influx but not inositol trisphosphate production in adrenal glomerulosa cell. FEBS Lett. 204:347–351

Lewis, D.L., Weight, F.F. & Luini, A. (1986) A guanine nucleotide-binding protein mediates the inhibition of voltage-dependent calcium current by somatostatin in a pituitary cell line. Proc. Natl. Acad. Sci. USA 83:9035–9039

Lux, H., Toselli, M. & Tokutomi, N. (1989) Transmitter-modulation of neuronal Ca channels. Pflüger's Arch. 413:R48 (Suppl.)

Martin, T., Bajjalieh, S., Lucas, D. & Kowalchyk, J. (1986) Thyrotropin-releasing hormone stimulation of polyphosphoinositide hydrolysis in GH_3 cell membranes is GTP dependent but insensitive to cholera or pertussis toxin. J. Biol. Chem. 261:10141–10149

Mullaney, I., Magee, A.I., Unson, C.G. & Milligan, G. (1988) Differential regulation of amounts of the guanine-nucleotide-binding proteins G_i and G_o in neuroblastoma × glioma hybrid cells in response to dibutyryl cyclic AMP. Biochem. J. 256:649–656

Mumby, S., Pang, I.-H., Gilman, A.G. & Sternweis, P.C. (1988) Chromatographic resolution and immunologic identification of the α_{40} and α_{41} subunits of the guanine nucleotide-binding regulatory proteins from bovine brain. J. Biol. Chem. 263:2020–2026

Naor, Z., Azrad, A., Limor, R., Zaku, H. & Lotan, M. (1986) Gonadotropin-releasing hormone activates a rapid Ca^{2+}-independent phosphodiester hydrolysis of polyphosphoinositides in pituitary gonadotrophs. J. Biol. Chem. 261:12506–12512

Offermanns, S., Schultz, G. & Rosenthal, W. (1989) Secretion-stimulating and secretion-inhibiting hormones stimulate high-affinity pertussis-toxin-sensitive GTPases in membranes of a pituitary cell line. Eur. J. Biochem. 180:283–287

Pfeuffer, T. (1977) GTP-binding proteins in membranes and the control of adenylate cyclase. J. Biol. Chem. 252:7224–7234

Rane, S.G. & Dunlap, K. (1986) Kinase C activator 1,2-oleoylacetyl glycerol attenuates voltage-dependent calcium currents in sensory neurons. Proc. Natl. Acad. Sci. USA 83:184–188

Rane, S.G., Walsh, M.P. & Dunlap, K. (1987) Norepinehrine inhibition of sensory neuron calcium current is blocked by a specific protein kinase C inhibitor. Soc. Neurosci. Abstr. 13:557

Rosenthal, W., Hescheler, J., Hinsch, K.-D., Spicher, K., Trautwein, W. & Schultz, G. (1988a) Cyclic AMP-independent, dual regulation of voltage-dependent Ca^{2+} channels by LHRH and somatostatin in a pituitary cell line. EMBO J. 7, 1627–1633

Rosenthal, W., Hescheler, J., Trautwein, W. & Schultz, G. (1988b) Receptor- and G-protein-mediated modulations of voltage-dependent calcium channels. Cold Spring Harbor Symp. Quant. Biol. 53:246–254

Schäfer, R., Christian, A.-L. & Schulz, I. (1988) Photoaffinity labeling with GTP-γ-azidoanilide of a cholera toxin-sensitive 40 kDa protein from pancreatic acinar cells. Biochem. Biophys. Res. Commun. 155:1051–1059

Schultz, G., Rosenthal, W., Trautwein, W. & Hescheler, J. (1990) Role of G-proteins in calcium channel modulations. Annu. Rev. Physiol. 52 (in press)

Tsunoo, A., Yoshii, M. & Narahashi, T. (1986) Block of calcium channels by enkephalin and somatostatin in neuroblastoma-glioma hybrid NG108–15 cells. Proc. Natl. Acad. Sci. USA 83:9832–9836

Ui, M., Katada, T., Murayama, T., Kurose, H., Yajima, M., Tamura, M., Nakamura, T. & Nogimori, K. (1984) Islet-activating protein, pertussis toxin: a specific uncoupler of receptor-mediated inhibition of adenylate cyclase. Adv. Cyclic Nucleotide Res. 17:145–151

van Dongen, A.M.J., Codina, J., Olate, J., Mattera, R., Joho, R., Birnbaumer, L. & Brown, A.M. (1988) Newly identified brain potassium channels are gated by the guanine nucleotide binding protein G_o. Science 242:1433–1437

Wanke, E., Ferroni, A., Malgaroli, A., Ambrosini, A., Pozzan, T. & Meldolesi, J. (1987) Activation of a muscarinic receptor selectively inhibits a rapidly inactivated Ca^{2+} current in rat sympathetic neurons. Proc. Natl. Acad. Sci. USA 84:4313–4317

Note added in proof:

Photo labeled 40 kDa proteins in membranes of GH_3 cells may represent α-subunits of pertussis toxin-sensitive *and* insensitive G-proteins.

Multiple Roles of G Proteins in Coupling of Receptors to Ionic Channels and Other Effectors

L. Birnbaumer[1], A. Yatani[1], J. Codina[1], A. VanDongen[1], R. Graf[1], R. Mattera[1], J. Sanford[1], and A.M. Brown[1]

1 Introduction

The central role of G proteins in coupling receptors to effector systems can be best illustrated by the contents of Table 1, which lists receptors that exert their actions by interacting with a G protein, according to the general schemes shown in Fig. 1. Like receptors, which are increasing in number rapidly, effectors affected by the activated forms of G proteins are also increasing, most notably through the discovery in 1986–1987 that ionic channels form part of the family of molecules regulated by G proteins. Using cell-free systems such as provided by excision of membrane patches from cells and incorporation of plasma membrane vesicles into lipid bilayers, it was shown that ionic channels are indeed regulated by activated G proteins. Some of these channels had long before been predicted to be under the control of G-protein-coupled receptors by means other than soluble second messengers. The mechanism by which G proteins regulate some of these channels is at the very least "membrane delimited" and independent of any phosphorylation event or of changes in cytoplasmic levels of second messengers such as cAMP, Ca^{2+}, or IP_3 and is very likely due to direct interaction of the G protein α-subunit and the channel proper (for review, see Brown and Birnbaumer 1988). Our group was prominent in providing some of the initial as well as subsequent supporting data for these conclusions (Brown and Birnbaumer 1988; Yatani et al. 1987a,b,c, 1988a,b,c; Codina et al. 1987a,b, 1988; Imoto et al. 1988; Kirsch et al. 1988). Another research group has claimed that ionic channels, specifically the heart muscarinic K^+ channel, now referred to as G_i-gated K^+ channel (see below), may be regulated by receptors, not via a specific G protein α-subunit, but via the $\beta\gamma$-subunits of G proteins (Logothetis et al. 1987). This has been corrected in part by Logothetis et al. (1988), who now recognize that α-subunits are active (Logothetis et al. 1988), and in our hands $\beta\gamma$-dimers antagonize rather than mimic effects of receptor stimulation (Codina et al. 1987a; Okabe et al. 1989). Table 2 lists regulations of ionic channels that were obtained under cell-free conditions with α-subunits and/or $\beta\gamma$-dimers. In addition there are reports on effects of GTPγS injection, of PTX treatment, and/or G-protein subunit injection into cells which are consistent with the possibility of further direct ionic channel regulations, but in which involvement of soluble second messengers has yet to be ruled in or out. These include inhibition with GTPγS or G proteins of the PTX-sensitive G_o-type of voltage-gated Ca^{2+} channels in chick (Holz et al. 1986) and rat (Scott and Dolphin 1986) dorsal root ganglion cells, in AtT-20

[1]Departments of Cell Biology and Physiology & Molecular Biophysics, Baylor College of Medicine, Houston, Texas 77030, USA

40. Colloquium Mosbach 1989
Molecular Mechanisms of Hormone Action

Table 1. Examples of receptors acting on cells via G proteins

Type of receptor	Membrane function/system affected[a]	Effect	Coupling protein involved	Examples of target cells(s)/organs
A. *Neurotransmitters*				
1. Adrenergic				
beta-1	AC	Stimulation	G_s	Heart, fat, symp. synapse
	Ca channel	Stimulation	G_s	Heart, skeletal muscle
beta-2	AC	Stimulation	G_s	Liver, lung
alpha-1	PhL C	Stimulation	G_{plc}	Smooth muscle, liver
	PhL A_2	Stimulation	G_{pla}	FRTL-1 cells
alpha-2a, -2b	AC	Inhibition	G_i	Platelet, fat (human)
	Ca channel	Closing	G_o (G_p)?	NG-108, Symp. presynapse
2. Dopamine				
D-1	AC	Stimulation	G_s	Caudate nucleus
D-2	AC	Inhibition	G_i	Pituitary lactotrophs
3. Acetylcholine				
muscarinic M_1	PhL C	Stimulation	G_{plc}	Pancreatic acinar cell
	K channel (M)	Closing	?	CNS, symp. ganglia
muscarinic M_2	AC	Inhibition	G_s	Heart
	K channel	Opening	G_k (G_i?)	Heart, CNS
	PhL C	Stimulation	G_p	Heart, transfected cells
4. $GABA_B$	Ca channel	Closing	G_o (G_p?)	Neuroblastoma N1E
	K channel	Opening	G_i (G_k?)	Sympathetic ganglia
5. Purinergic Pl				
adenosine A-1 or Ri	AC	Inhibition	G_i	Pituitary, CNS, heart
	K channel	Opening	G_k (G_i?)	Heart
adenosine A-2 or Ra	AC	Stimulation	G_s	Fat, kidney, CNS
6. Purinergic P_{2X} and P_{2Y}	PhL C (PIP_2)	Stimulation	G_{plc}	Turkey erythrocytes
	PhL C (PC)	Stimulation	G-(?)	Liver
7. Serotonin (5HT)				
S-1a (5HT-1a)	AC	Inhibition	G_i	Pyramidal cells
	K channel	Opening	(G_i/G_o?)	Pyramidal cells
S-1c (5HT-1c)	PhL C	Stimulation	G_{plc}	Aplysia
S-2 (5HT-2)	AC	Stimulation	G_s	Skeletal muscle

8. Histamine				
H-1	PhL C	Stimulation	G_{plc}	Smooth muscle, macrophages
	PhL A_2 (?)	Stimulation	G_{pla}	
H-2	AC	Stimulation	G_s	Heart
H-3	AC	Inhibition	G_i	Presynaptic CNS, lung; mast
B. *Peptide hormones*				
1. Pituitary				
adrenocorticotropin (ACTH)	AC	Stimulation	G_s	Fasciculata, glomerulosa
opioid (mu, kappa, delta)	AC	Inhibition	G_i	NG-108
	Ca channel	Closing	G_o (G_p?)	NG-108
luteinizing hormone (LH)	AC	Stimulation	G_s	Granulosa, luteal, Leydig
follicle stimulating hormone (FSH)	AC	Stimulation	G_s	Granulosa
thyrotropin (TSH)	AC	Stimulation	G_s	Thyroid, FRTL-5
	PhL?	Stimulation	G_p(?)	Thyroid
melanocyte stimulating hormone (MSH)	AC	Stimulation	G_s	Melanocytes
2. Hypothalamic				
corticotropin releasing hormone (CRF)	AC	Stimulation	G_s	Corticotroph, hypothalamus
growth hormone releasing hormone (GRF)	AC	Stimulation	G_s	Somatotroph
gonadotropin releasing hormone (GnRH)	PhL A_2	Stimulation	G_{pla}	Gonadotroph
	PhL C	Stimulation	G_{plc}	Gonadotroph
	Ca channel	Opening	G_i-type	GH_3
thyrotropin releasing hormone (TRH)	PhL C	Stimulation	G_{plc}	Lactotroph, thyrotroph
	AC	Inhibition	G_i	GH_4C_1
somatostatin (SST or SRIF)	AC	Inhibition	G_i	Pit. cells, endocr. pancr.
	K channel	Opening	G (G_i?)	Pit. cells, endocr. pancr.
	Ca channel	Closing	?	Pit. cells
vasopressin				
V-1a (vasopressor, glycogenolytic)	PhL C	Stimulation	G_{plc}	Smooth muscle, liver, CNS
	AC	Inhibition	G_i	Liver
V-1b (pituitary)	PhL C	Stimulation	G_{plc}	Pituitary
V-2 (antidiuretic)	AC	Stimulation	G_s	Distal and collecting tubule
Oxytocin	PhL C	Stimulation	G_{plc}	Uterus, CNS
3. Other hormones				
chorionic gonadotropin	AC	Stimulation	G_s	Granulosa, luteal, Leydig
glucagon	AC	Stimulation	G_s	Liver, fat, heart
	Ca pump	Inhibition	G_s(?)	Liver, heart (?)

Table 1. *continued*

Type of receptor	Membrane function/system affected[a]	Effect	Coupling protein involved	Examples of target cells(s)/organs
	PhL C	Stimulation	?	Liver
cholecystokinin (CCK)	PhL C	Stimulation	G_{plc}	Pancreatic acini
secretin	AC	Stimulation	G_s	Pancreatic duct, fat
vasoactive intestinal peptide (VIP)	AC	Stimulation	G_s	Pancreatic duct, CNS
	PhL C	Stimulation	G_{plc}	Sensory ganglia, CNS
angiotensin II	PhL C	Stimulation	G_{plc}	Liver, glomerulosa cells
	AC	Inhibition	G_i	Liver, glomerulosa cells
	Ca channel	Stimulation	G_i-type	Yl adrenal cells
parathyroid hormone	AC	Stimulation	G_s	Kidney, bone
calcitonin	AC	Stimulation	G_s	Bone
C. *Other regulatory factors*				
1. Chemoattractant (fMet-Leu-Phe or fMLP)	PhL C	Stimulation	G_{plc}	Neutrophils
2. Thrombin	PhL C	Stimulation	G_{plc}	Platelets, fibroblasts
	AC	Inhibition	G_i	Platelets
3. Bombesin	PhL C	Stimulation	G_{plc}	Fibroblasts
4. IgE	PhL C	Stimulation	G_{plc}	Mast cells
5. Bradykinin	PhL C	Stimulation	G_{plc}	Lung, fibroblasts, NG-108
	PhL A_2	Stimulation	G_{pla}	Fibroblasts, endothel. cells
	K channel	Stimulation	G_k (G_i?)	NG-108
	AC	Inhibition	G_i	NG-108
6. Neurokinin/tachykinin				
7. Neuropeptide Y	K channels	Stimulation	G_k	Heart
	Ca channels	Inhibition	G_o	Sensory ganglia
8. Tumor necrosis factor (TNF)	?	?	?	Monocytes
9. Colony stimulating factor (CSF-1)	?	?	?	Monocytes

D. *Prostanoids*				
1. Prostaglandin E_1, E_2	AC	Inhibition	G_i	Fat, kidney
2. Prostacyclin (PGI_2, PGE_1, PGE_2)	AC	Stimulation	G_s	Luteal cells, endothel., kidney
3. Tromboxanes	PhL C	Stimulation	G_{plc}	Platelets
4. Platelet activating factor (PAF)	PhL C	Stimulation	G_{plc}	Platelets
5. Leukotriene D_4, C_4	PhL A_2	Stimulation	G_{pla}	Endothelial cells
E. *Sensory*				
1. Light				
rhodopsins	cGMP-PDE	Stimulation	$Tr(G_{t-r})$	Retinal rod cells (night)
opsins (blue, red, and green)	cGMP-PDE	Stimulation	$Tc(G_{t-c})$	Retinal cone cells (color)
2. Olfactory signals	AC	Stimulation	G_{olf}	Olfactory cilia
	PhL's?	Stimulation	G_p?	
3. Taste signals	AC	Stimulation	G_s	Taste epithelium
	PhL's	Stimulation	G_p?	

[a] AC: adenyl cyclase; PhL C: unless denoted otherwise, phospholipase C with specificity for phosphatidylinositol bisphosphate; PhL A_2: phospholipase A_2 (substrate specificity unknown); PIP_2: phosphatidylinositol bisphosphate; PC: phosphatidylcholine.

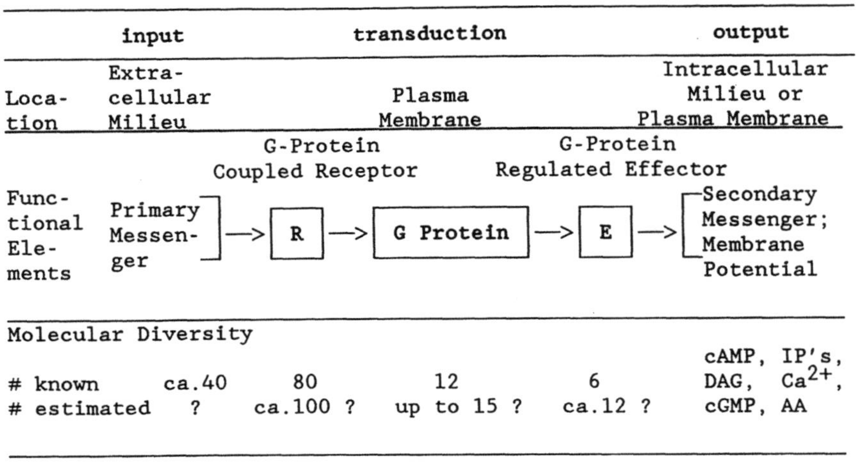

Fig. 1. Flow of information through G protein-dependent signal transduction systems as found in vertebrates

(Lewis et al. 1986) and NG108–15 (Hescheler et al. 1987) cells and in rat superior cervical ganglion cells (Ewald et al. 1988), and the stimulation of K^+ channels in adrenal Y1 and GH_3 cells (Hescheler et al. 1988; Rosenthal et al. 1988).

The case of cell-free regulation of dihydropyridine (DHP)-sensitive Ca^{2+} channels by G_s is of special interest; it was unexpected for of two reasons: one because DHP-sensitive Ca^{2+} channels had been shown to be stimulated upon phosphorylation by the catalytic unit of cAMP-dependent protein kinase (Kameyama et al. 1986), indicating that, as summarized in Fig. 2, nature uses dual pathways to regulate a single function, one fast and membrane-delimited, the other slower with a longer life span; the other, because it had been thought that G proteins might be "monogamous", i.e., specific for single effector functions, and here we were faced with proof for multifunctionality in G-protein actions.

These advances were all the result of a multidisciplinary approach to the problem of signal transduction by G proteins which brought together classical biochemistry, sophisticated single channel recordings, and modern molecular biology. The background experiments, especially those of Nargeot et al. (1983), Soejima and Noma (1984), Pfaffinger et al. (1985), and Breitweiser and Szabo (1985), which led to the discovery that ionic channels are effector systems of G proteins akin to adenylyl cyclase and cGMP-phosphodiesterase as well as the initial experiments showing effects of purified G proteins and protein subunits on channels in excised membrane patches, were reviewed in Brown and Birnbaumer (1988). This chapter will center on some of the more recent results from our laboratories dealing mostly, but not exclusively, with a pertussis toxin-sensitive G_i-type family of G proteins, their effects of ionic channels, our efforts in assigning defined functions to individual G proteins as they are known from biochemical and molecular cloning studies, and some speculations that follow from the results obtained as to why it is that G proteins dissociate and how signal transduction pathways are set up.

Table 2. Ionic channels regulated by G proteins under cell-free conditions in a membrane-delimited ("direct") manner

Species	Tissue/cell line	Method	Channel properties: g	Tau	Rectification	G protein type	Effect of G protein on channel kinetics	Ref[a]	Agonist(s)
				(ms)					
A. K^+ Channels									
Guinea pig	Heart atria	I-O patch	40 pS*	1.2	Inward	G_i-3	Increase in Po	1	Ach, Ado
		"				α_i-3(α_{40})	" " "	2–4	
		"				α_i–1,–2,–3, r α_i	" " "	5,6	
		"				$\alpha_{36/35}\ \gamma_G$	Inhibition of gating by G_k	7	
		"				$\beta_{35}\ \gamma_G$	" " " " "	7	
		"				$\beta_{36}\ \gamma_T$	" " " " "	7	
Chicken	Embryonic atria	I-O patch	40 pS	0.9	Inward	$\beta_{36/35}\ \gamma_G$	Increase in Po	8	Ach, Ado
		"				α_i-3, α_{40}	" " "	3,4,9	
		"				α_{39}	" " "	9	
		"				α_{41} (brain, liver)	No effect	9	
Rat	Neonatal atria	I-O patch	40 pS*	1.0	Inward	α_i-3	Increase in Po	3	Ach, Ado,
	GH_3 pit. cell	I-O patch	55 pS*	1.5	?	G_i-3	" " "	10	ACh, SST
						α_i-3	" " "	11	
	Hippocampus	I-O patch	13 ps		No	G_o	" " "	12	? (5HT,
		"	38 pS		No	rα_o	" " "	12	GABA
		"	38 pS		Inward	G_o	" " "	12	Norepi)
		"	55 pS		No	G_o	" " "	12	
B. ATP-sensitive K^+ channels									
Rabbit	Sk. muscle T-tub.	Lipid bilayer				α_i-3		13	ATP
Rat	RINm5F	I-O patch				α_i-3		14	ATP (glucose)
C. Voltage-gated Ca^{2+} channels, DHP-sensitive									
Guinea pig	Heart ventricle	I-O patch				G_s, α_s	Increased survival	15	Epi, Norep
		Lipid bilayer	25 pS			G_s	Increased Po	15	
		" "	25 pS			G_s, α_s	Increased open time Decreased inactivation time	16	

Table 2. *Cont.*

Species	Tissue/cell line	Method	Channel properties: g	Tau	Rectification	G protein type	Effect of G protein on channel kinetics	Ref[a]	Agonist(s)
Rabbit	Sk. muscle T-tub.	Lipid bilayer	10 pS			G_s, α_s		17	Epi. Norep
		" "	10 pS			$r\alpha_s$	Increased NP	18	
C. Other									
a. Voltage-gated Na^+ *channel, TTX-sensitive*									
Guinea pig	Heart	I-O patch				α_s		19	Epi. Norep
b. Cation amiloride- and PTX-sensitive cation channel (apical membrane)									
Rat	Collecting tubule	I-O patch				α_i-3		20	Any?

[a] *References:* 1. Yatani et al. (1987a); 2. Codina et al. (1987a); 3. Kirsch et al. (1988); 4. Cerbai et al. (1988); 5. Yatani et al. (1988c); 6. Mattera et al. (1989a); 7. Okabe et al. (1989); 8. Logothetis et al. (1987); 9. Logothetis et al. (1988); 10. Yatani et al. (1987b); 11. Codina et al. (1987b); 12. VanDongen et al. (1988); 13. Parent and Coronado (1988); 14. Ribalet et al. (1989); 15. Yatani et al. (1987c); 16. Imoto et al. (1988); 17. Yatani et al. (1988a); 18. Mattera et al. (1989b); 19. Schubert et al. (1989); 20. Light et al. (1989).

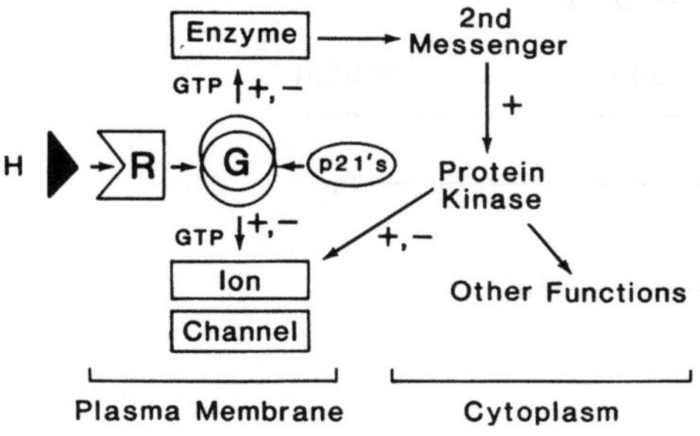

Fig. 2. General diagram of involvement of both direct G-protein regulation and indirect G-protein regulation of a single effector function as well as of the dual effect G proteins may have as intramembrane signal transducers

2 Primary Structure of G Proteins

The primary structure of G proteins has been reviewed recently by Lochrie and Simon (1988), as well as by us (Birnbaumer et al. 1989). Briefly, G proteins are heterotrimers formed of α-, β-, and γ-subunits. At the time of this writing 12 α-subunits, encoded in 9 genes, 2 β-subunits, and at least 2 γ-subunits are known. Of these, all the α-, the two β-, and one of the γ-subunits have been cloned (Fig. 3) and their amino acid sequence deduced (e.g., Fig. 4). β- and γ-subunits form dimers which may be of two types, $\beta_{36}\gamma$ and $\beta_{35}\gamma$, if a cell expresses only one type of γ-subunit, or of four types if it expresses two. To our knowledge, α-subunits, when combining with $\beta\gamma$-dimers to form holo-G proteins, do not distinguish among $\beta\gamma$-dimers (Fig. 5A). This is not to say that all tissues have the same complement of $\beta\gamma$-dimers. Quite the contrary, as illustrated by the differing β_{35}/β_{36} ratio found, for example, in human placenta, human erythrocytes, bovine brain, and bovine retinal rod cells (Fig. 5B). α-Subunits bind GTP, hydrolyze GTP, dissociate from the $\beta\gamma$-dimer on activation by GTP analogs, and/or the $NaAlF_4$/GDP, and, with few exceptions, are substrates for ADP-ribosylation by cholera and/or pertussis toxin (CTX and PTX). Studies with transducin-α identified an arginine at the approximate center of the molecules as the aminoacid ADP-ribosylated by CTX (VanDop et al. 1984) and a cysteine at position –4 from the carboxyl terminal end as the site of ADP-ribosylation by PTX (West et al. 1985). Figure 6 shows typical SDS-PAGE patterns obtained when membranes from various sources are ADP-ribosylated with CTX and PTX using $[^{32}P]NAD^+$ and electrophoresed through a standard 10% acrylamide gel.

Not all G proteins have been either purified or cloned. This is best illustrated on analyzing pertussis toxin labeling patterns by high resolution urea gradient/SDS-PAGE (Scherer et al. 1987) instead of the standard SDS-PAGE (Fig. 7). Thus, even though molecular cloning has revealed the existence of many more G proteins than anticipated from purification studies, especially in the fields of G_s and G_i, for what was originally thought to be two forms of α_s (Robishaw et al. 1986; Mattera et al. 1986) turned out to be four splice variants (Bray et al. 1986; Kozasa et al. 1988) and what was originally thought of as one α_i turned out to be three α_i's encoded in three independent genes, as deduced from cDNA cloning (Suki et al. 1987; Jones and Reed 1987; Beals et al. 1987) and gene sequencing (Itoh et al. 1988), there appear to exist still more G protein α-subunits when viewed by PTX labeling for which cDNA cloneh have

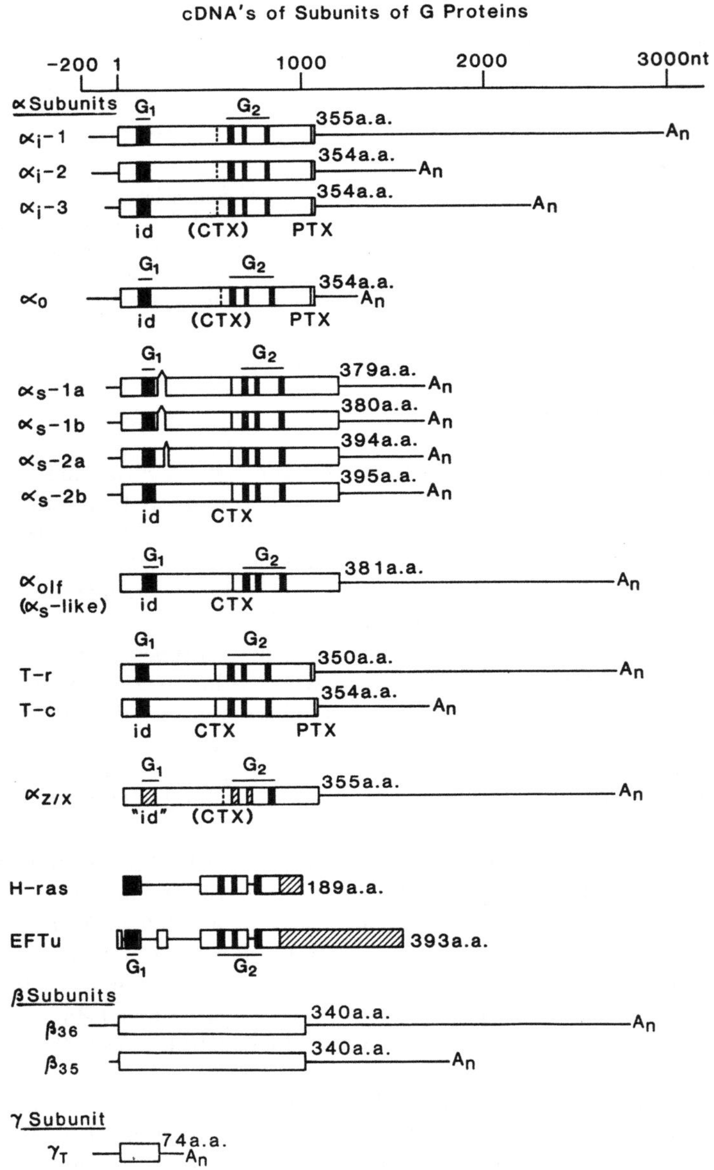
cDNA's of Subunits of G Proteins
-200
1
1000
2000
3000nt
∝Subunits
G1
G2
∝i-1
355a.a.
An
∝i-2
354a.a.
An
∝i-3
354a.a.
An
id
(CTX)
PTX
∝o
354a.a.
An
∝s-1a
379a.a.
∝s-1b
380a.a.
∝s-2a
394a.a.
∝s-2b
395a.a.
CTX
∝olf
(∝s-like)
381a.a.
T-r
350a.a.
T-c
354a.a.
∝z/x
355a.a.
"id"
H-ras
189a.a.
EFTu
393a.a.
βSubunits
β36
340a.a.
β35
340a.a.
γ Subunit
γT
74a.a.

◄

Fig. 3. Diagram of cloned cDNAs encoding α-, β-, and γ-subunits of G proteins. *Open boxes* represent the open reading frames or coding sequences and *lines* represent 5′ and 3′ untranslated sequences which may be incomplete. *Black boxes* within the open reading frames of α-subunits denote sequences highly homologous to those known in bacterial elongation factor TU to be involved in GTP binding and hydrolysis. Sequences homologous to these are present also in the *ras* molecules. *Hatched areas* in EF TU and *ras* are non-homologous to α-subunits of hetrotrimeric G proteins (Halliday 1985; Jurnak et al. 1986; Masters et al. 1986) The position of amino acids ADP-ribosylated by CTX and PTX are indicated. *i.d.*, identity box: a stretch of 18 invariant amino acids in all (*black*) except $\alpha_{Z/X}$, which differs not only within the *i.d.* box but also in two of the other three regions involved in GTP binding (*hatched*). The scale is in nucleotides (*nt*). At present, the following α-subunits are known to exist. Four splice variants of α_s (Harris et al. 1985; Nukada et al. 1986a; Robishaw et al. 1986; Itoh et al. 1986; Mattera et al. 1986; Sullivan et al. 1986; Bray et al. 1986; Jones and Reed 1987) encoded in a single gene (Kozawa et al. 1988); they stimulate adenylyl cyclase and at least two Ca^{2+} channels of the DHP-sensitive type and a TTX-sensitive Na^+ channel. There are three α_i genes (Nukada et al. 1986a,b; Itoh et al. 1986; Sullivan et al. 1986; Jones and Reed 1987; Didsbury et al. 1987; Bray et al. 1987; Didsbury and Snyderman 1987; Suki et al. 1987; Beals et al. 1987; Codina et al. 1988; Kim et al. 1988; Itoh et al. 1988) shown all to stimulate atrial K^+ channels. They may be multifunctional. Even though biochemically it is possible to purify at least two types of G_o proteins (Fig. 9), only one type of cDNA has been cloned (Itoh et al. 1986; Jones and Reed 1987; VanMeurs et al. 1987; VanDongen et al. 1988).

There are two transducin genes (Tanabe et al. 1985; Medynski et al. 1985; Yatsunami and Khorana 1985; Lochrie et al. 1985), one is expressed in rod cells and the other in cone cells of the retina (Grunwald et al. 1986; Van Veen et al. 1986; Lerea et al. 1986). Transducin has also been found in pineal cells, but has not been characterized. There is in addition an α_s-type α-subunit (α_{olf}) expressed only in neuro-olfactory epithelium (Jones and Reed 1988). An α-subunit differing in the identity box and lacking a Cys for ADP-ribosylation by PTX has also been cloned (Fong et al. 1988; Matsuoka et al. 1988).

Two β-subunits have been cloned. One, β_{36} (Sugimoto et al. 1985; Codina et al. 1986; Fong et al. 1986) is expressed in rod cells as well as most other tissues except perhaps placenta, the other, β_{35} (Gao et al. 1987; Fong et al. 1987) is expressed in most tissues, including placenta where it is the predominant β-subunit (Fig. 3), but not in rod cells. One γ-subunit, expressed only in rod cells (Gierschik et al. 1985) has been cloned (Yatsunami et al. 1985; Hurley et al. 1984). That (those) γ-subunit(s) expressed in all other tissues has (have) not been cloned as yet. α-Subunits have also been cloned from lower eukaryotes including yeast (Dietzel and Kurjan 1987; Nakafuku et al. 1987, 1988) and *Drosophila* (Provost et al. 1988). Homologous forms of β and γ have been cloned from yeast (Whiteway et al. 1989) as well as a β-subunit from *Drosophila* (Yarfitz et al. 1988)

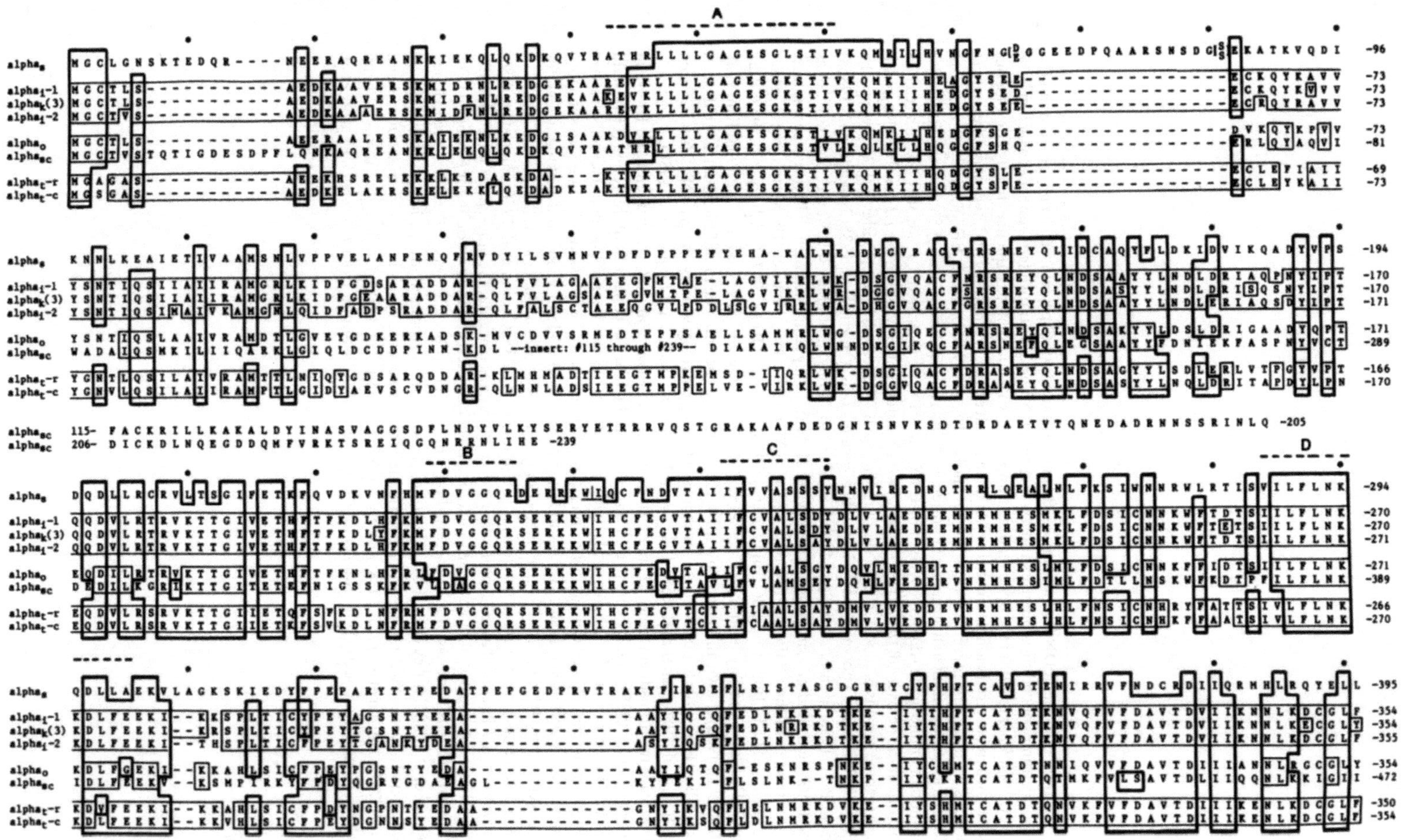

Fig. 4. Deduced amino acid composition and comparative alignment of human α_s, human forms of α_i-1, -2, and -3; bovine α_o (not known if type 1 or 2), bovine α_t (rod) and yeast GPA-1, a highly homologous α-subunit involved in the pheromone signal transduction pathway (Miyajima et al. 1987; Kurjan 1989). Amino acid identities are *boxed*

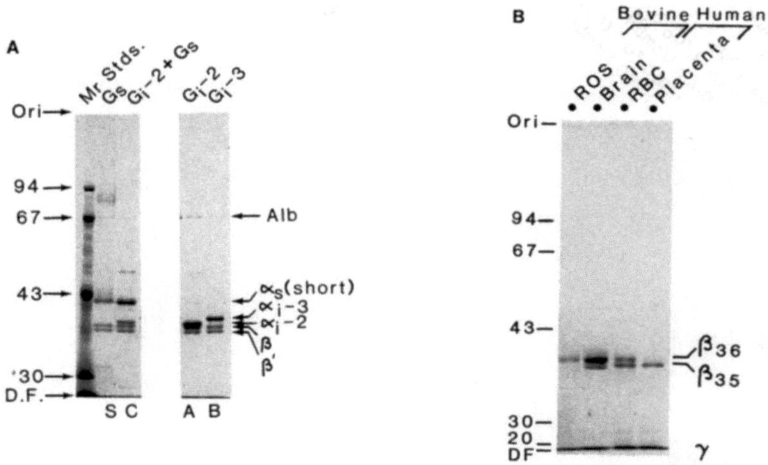

Fig. 5A,B. Urea gradient/SDS-PAGE analysis of human erythrocyte G_i-2, G_i-3, and G_s (**A**) and of various $\beta\gamma$-preparations (**B**). Human erythrocyte proteins for **A** were obtained by DEAE-Toyopearl purification (Okabe et al. 1989). Note the variable ratio between the two β-subunits in different tissues. Placental $\beta\gamma$-dimer was a gift from Dr. Tony Evans (Genentech, Inc.); rod outer segment (*ROS*) $\beta\gamma$ was a gift from Dr. K.J. Ho, University of Illinois School of Medicine, Chicago. Photographs of Coomassie blue stained gels are shown

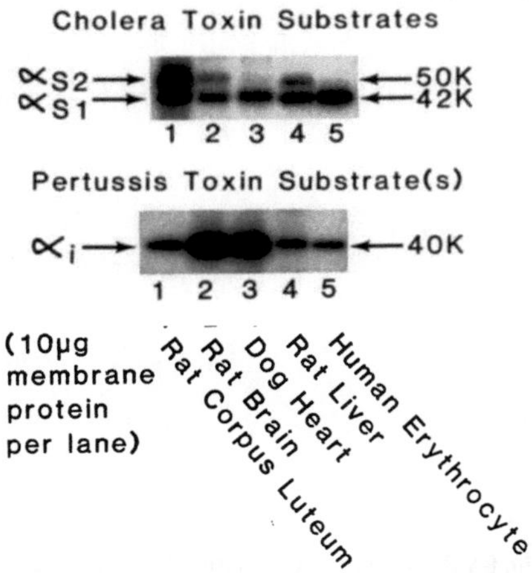

Fig. 6. G proteins as visualized by toxin labeling analyzed by 10% SDS-PAGE followed by autoradiography. *Top* Labeling of membrane proteins with CTX and [^{32}P]NAD (21-h autoradiogram); *bottom* labeling of membrane proteins with PTX and [^{32}P]NAD (4-h autoradiogram). Photographs of the regions of molecules between 35 and 60 kDa are shown. Heavy bands obtained with PTX in heart and brain are in fact at least two G_i and two G_o molecules which can be resolved by urea-gradient/SDS-PAGE (Scherer et al. 1987; Figs. 7 and 8)

not yet been found. Not all of these "additional" G proteins are of the quantitatively "minor" class, as exemplified by the existence of a novel type of G_o (G_{o2}), which by labeling (Scherer et al. 1987) and purification (Codina and Birnbaumer unpublished) exceeds G_{i1} in brain. The relative migration of the α-subunits of the three G_i's and the two G_o's, is illustrated in Fig. 8.

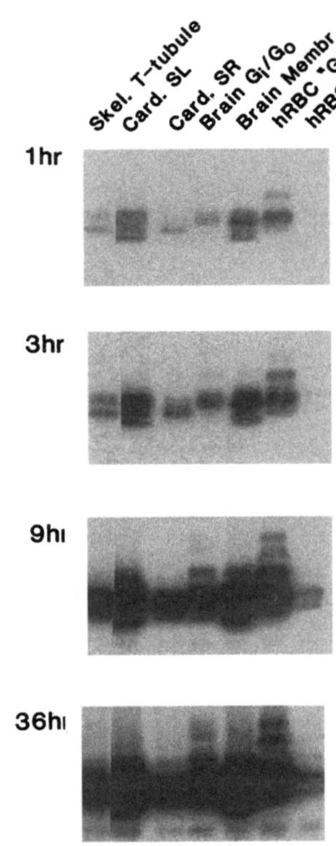

Fig. 7. Membranes and "purified" proteins contain multiple PTX substrates. PTX-labeled membranes (1 μg/lane) or purified bovine brain mixture of G/G_o and human erythrocyte G_i (20 ng/lane) were separated by SDS-PAGE in 8% polyacrylamide gels in the presence of a 4 to 8 M urea gradient, the gel slabs were fixed, stained with Coomassie blue and destained (not shown) and then autoradiographed for the times indicated. Only the section of the autoradiograms with molecules between 35 and 45 kDa are shown

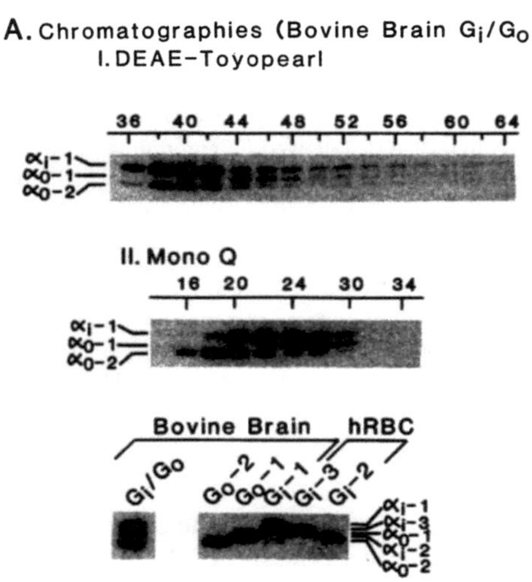

Fig. 8A,B. Migration of α-subunits of purified G proteins after ADP-ribosylation with pertussis toxin. **A** Aliquots of native G proteins purified by FPLC over MonoQ or DEAE-Toyopearl were ADP-ribosylated and separated by SDS-Urea gradient/9% PAGE (Scherer et al. 1987). **B** Analysis of purified pools of PTX substrate

3 Functional Studies

3.1 Combined Use of Natural and Recombinant α-Subunits Made in Bacteria to Define Their Functions

The abundance at which different α_s and α_i molecules are expressed varies from tissue to tissue raising the question as to whether functional differences are associated with the structural differences, or whether the G_s's and, respectively, the G_i's should be thought of merely as isoforms. Although the final word on these questions is not yet in, we have during this last year developed the method(s) required to answer them. Thus, we succeeded in purifying two types of G_i from human erythrocytes (hRBCs) (Fig. 5) and a third from bovine brain, from which we also purified the two forms of G_o (shown in their ADP-ribosylated form in Fig. 8). This allowed us to test for potential differential biological functions. We also expressed biologically active forms of the α-subunits in bacteria, designated as recombinant α-subunits, so that we could predict/confirm biological functions of various cloned and/or purified G protein α-subunits.

We have not yet carried out all of the studies. However, we were able to determine first of all that both the natural purified and the recombinant forms of α_{i3}, α_{i1}, and α_{i2} all stimulate K^+ channels (Mattera et al. 1989a; Yatani et al. 1988c; Figs. 9 and 10). We failed to observe significant differences between types 1, 2, and 3 α_i molecules, even though the recombinant forms all had potencies between 30- and 50-fold lower than their natural counterparts (Yatani et al. 1988c; Fig. 11). Thus, with respect to atrial muscarinic K^+ channels, G_{i1}, G_{i2}, and G_{i3} must be considered iso-G proteins. Studies are currently in progress in atrial membrane patches in which the endogenous G_k has been uncoupled from receptors by treatment with PTX (Yatani et al. 1987b; Brown and Birnbaumer 1988) to determine whether muscarinic and/or P_2 purinergic receptors exhibit selectivity for interaction with one or another of the G_i proteins.

Second, by testing the effect of recombinant α_s, we were able to determine that the stimulation of Ca^{2+} channels obtained with purified hRBC G_s (Yatani et al. 1987c, 1988a; Imoto et al. 1988) is indeed mediated by G_s as opposed to being due to a contaminant (Fig. 12). Further, in collaboration with M. Graziano and A. Gilman we tested the recombinant forms of three of the possible four splice variants of α_s for their Ca^{2+} channel stimulatory activity and found that they all do so with indistinguishable potency and efficacy (Mattera et al. 1989b). As was the case with recombinant α_i-subunits, the recombinant α_s-subunits also displayed a potency that was markedly (ca. 20-fold) reduced with respect to that of native human erythrocyte α_s. This applied not only to Ca^{2+} channel stimulation, but equally to adenylyl cyclase stimulation. Very likely bacteria fail to carry out a critical posttranslational modification that exists only in eukaryotic cells, or alternatively, modify the α-subunit in a manner that eukaryotic cells do not.

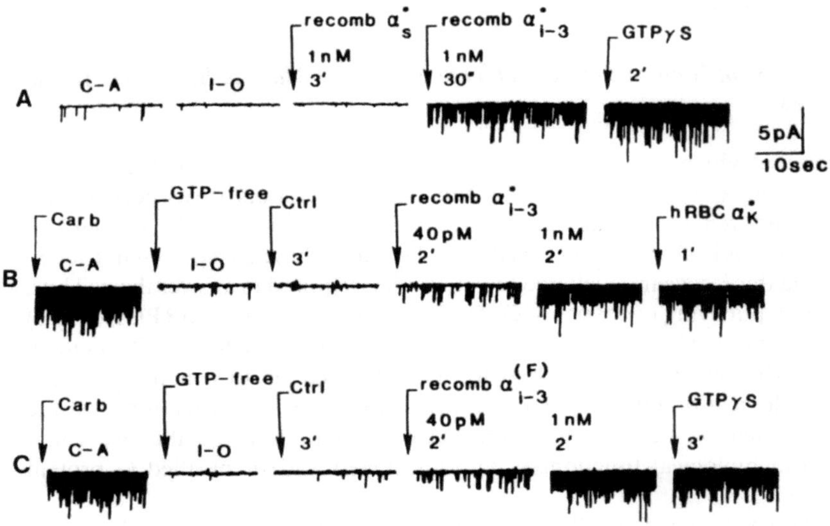

Fig. 9A-C. Stimulation of the atrial G protein-gated K$^+$ channel by recombinant α_i-3. Recombinant α_i-3 and α_s-subunits encoded in their respective cDNAs inserted into pT7–7 vector were expressed in *E. coli* by the one plasmid/one phage method of Tabor and Richardson (Tabor and Richardson 1985; Tabor 1987; Tabor et al. 1987), the cells were lysed and the recombinant α-subunits were activated and purified partially by chromatography over DEAE-Sephacel (0 to 400 nM NaCl gradient). The partially purified recombinant GTPγS-activated α-subunits were then added to inside-out membrane patches of guinea pig atria to test for possible K$^+$ channel stimulatory effects. Experiment **A**: Recombinant α_s does not but recombinant α-3 does stimulate the atrial K$^+$ channel. Experiments **B** and **C**: Effects of two different doses of GTPγS-activated or fluoride-activated recombinant α_i-3. In these and repeat experiments recombinant α_i-3 was about 3 to 10% as active as purified human erythrocyte GTPγS-activated G_k

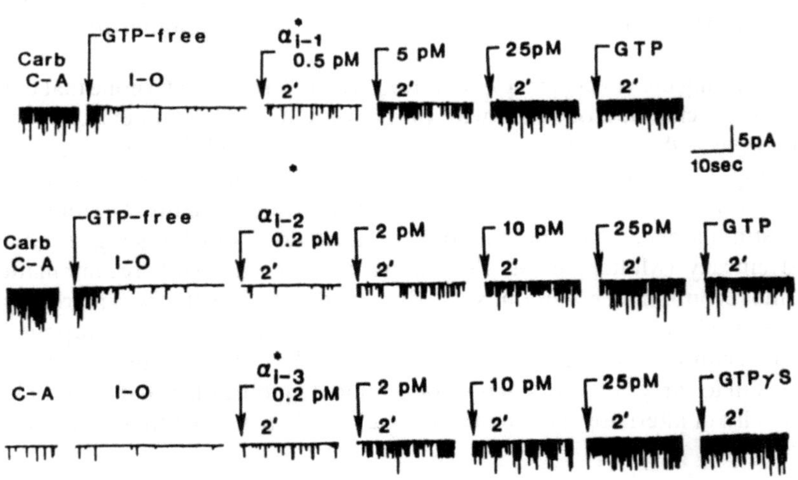

Fig. 10. Stimulation of single channel K$^+$ currents in adult guinea pig atrial membrane patches by all three types of native GTPγS-activated (*) α_i-subunits. G_i-2 and G_i-3 (Fig. 4) were purified by DEAE-Toyopearl, and G_i-1 was purified by Mono-Q FPLC. The purified proteins were activated with GTPγS and their α-subunits separated from $\beta\gamma$-dimers by DEAE-Sephacel chromatography (Codina et al. 1987a,b). Bovine brain G_i-1 was a generous gift from Drs. Elena Padrell and Ravi Iyengar (Mount Sinai School of Medicine)

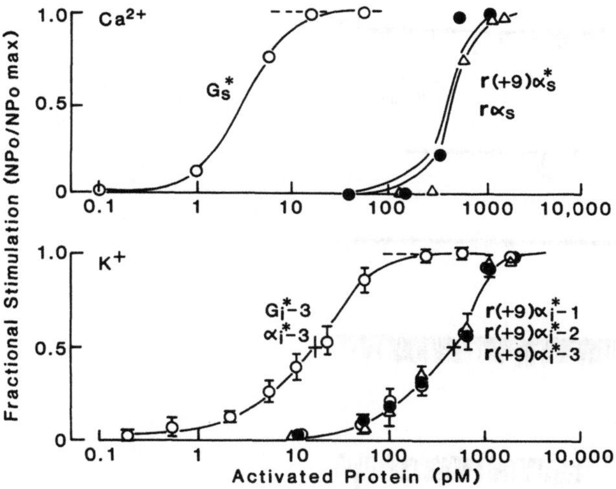

Fig. 11. Dose-response relationships for effects of native and recombinant α-subunits on skeletal muscle Ca^{2+} channel activity (*upper panel*) and atrial K^+ channel activity (*lower panel*). *Upper panel:* G_s^*, GTPγS-activated human erythrocyte G_s (unknown mix of short forms) plus an equimolar concentration of βγ-dimers r(+9)α_s^*, GTPγS-activated recombinant α_s biosynthesized in bacteria from α_s cDNA (short minus Ser version) placed into the PT7-vector and having a nine amino acid extension at its amino terminal end (Mattera et al. 1989a); rα_s^*, recombinant α_s molecule prepared by Drs. Graziano and Gilman, in Dallas, using a pT5 expression vector that yields α_s-subunits (long plus Ser version) without added amino acids. For details of synthesis, see Graziano et al. (1987) and of effects of recombinant α_s subunits on Ca^{2+} currents, see Mattera et al. (1989b). *Lower panel:* α_i^*-3, GTPγS-activated α-subunit of human erythrocyte G_i-3 after separation from βγ-dimers by DEAE-Toyopearl chromatography; G_i^*, GTPγS-activated human erythrocyte G_i-3 before DEAE-Toyopearl chromatography; the recombinant GTPγS-activated forms of the three types of α_i made in bacteria with the help of the pT7 expression vector are denoted as r(+9)α_i^*-x. Adapted from Yatani et al. (1988c)

3.2 An Alternative Form of Recombinant α-Subunit: Expression in Sf9 Insect Cells Under the Control of Baculoviral Polyhedrin Promoter

Following the strategy pioneered by Luckov and colleagues (e.g., Luckov and Summers 1988; Fig. 13) we prepared recombinant viral particles of the baculovirus *A. californica*, in which the open reading frames of α-subunit cDNA's had been placed under the control of the polyhedrin promoter, and expressed the α-subunits in *S. frugiperda-9* (Sf9) ovary cells grown in suspension culture. As illustrated in Fig. 14, this led to the accumulation of up to 20–25% of the total cell protein as α-subunit protein, of which 15 (α_s-subunits) to 30% (α_{i3}-subunits) was recovered after homogenization as soluble protein in 30 000 to 100 000 × g supernatants. A single column chromatography led to isolation of an α_{i3} preparation ca. 40% pure. Although we have not yet tested this preparation for K^+ channel stimulatory activity, a preliminary characterization has shown it to interact with βγ-dimers to form trimers susceptible to ADP-ribosylation by PTX about 50% as well as natural human erythrocyte G_{i3} and to bind 0.72–0.76 mol [^{32}] GTPγS per mole of baculoviral α_{i3} which is retained on the molecule under the influence of Mg^{2+} (Fig. 15).

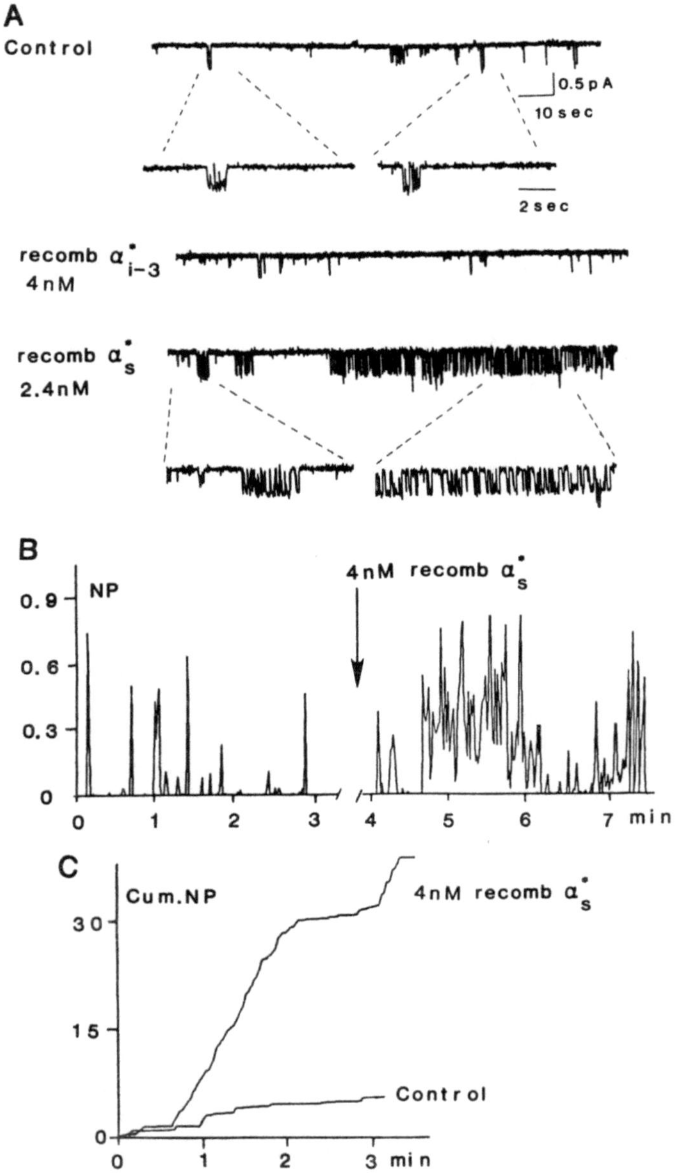

Fig. 12A-C. Stimulation of skeletal muscle T-tubule dihydropyridine-sensitive Ca^{2+} channel incorporated into lipid bilayers by the short form of recombinant α_s. Recombinant α_s (Asp^{71} version) and recombinant α_i-3 were prepared in their GTPγS-activated form and partially purified by DEAE-Sephacel chromatography, except that the two-plasmid approach was used to express the recombinant proteins (Tabor and Richardson 1985; Tabor 1987; Tabor et al. 1987). **A** Single channel Ba^{2+} currents (*trans* to *cis*) before and after addition first of GTPγS-activated recombinant α_i-3 and then of GTPγS-activated recombinant α_s. **B, C** NPo diaries of Ca^{2+} channel activities before and after addition of GTPγS-activated recombinant α_s to the *cis* chamber (**B**) and cumulative NPo values as a function of time of the same experiment (**C**)

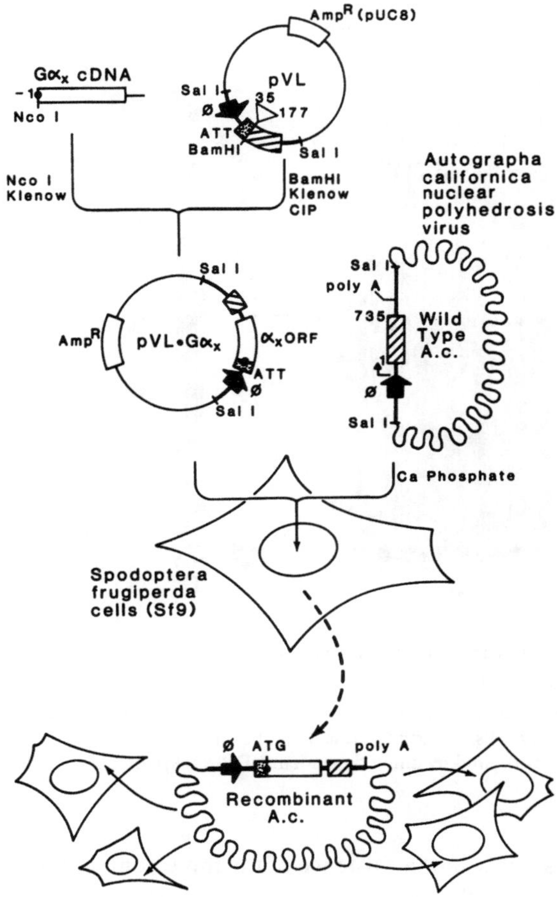

Fig. 13. Strategy for expression of G-protein α-subunits by the baculovirus-based expression system. The open reading frame of protein to be expressed is subcloned into plasmid vector pVL, which contains a modified polyhedrin gene of the baculovirus *A. californica*, in such manner that the coding region of the polyhedrin gene is replaced with that of the protein to be expressed. The recombinant plasmid DNA is cotransfected with wild-type baculovirus DNA into permissive cells of the silk worm *S. frugiperda* (Sf9 cells), and recombinant viral particles are isolated and plaque purified. Bulk quantities of sf9 cells are then infected with recombinant baculovirus for production of recombinant protein instead of polyhedrin. (For details, see Luckov and Summers 1988)

3.3 Gating of Ionic Channels as a Tool To Discover New Roles for G Proteins: Effects of G_o on Neuronal K^+ Channels

One of the properties of the "muscarinic" K^+ channels is that they are essentially silent in the absence of stimulation by an activated G protein (G_k). That is, in the absence of activated G protein their Po is close to zero. The possibility existed that not only G_i proteins regulate K^+ channels but also the structurally closely related G_o. Since nerve tissue is rich in G_o, central nervous system neurons, specifically hippocampal pyramidal cells, were placed into culture and studied for potential presence of both G_i- and G_o-gated K^+ channels. Although these studies are still in progress (VanDongen, Birnbaumer, Brown, unpublished), the initial findings with highly purified bovine brain GTPγS-activated G_{o1} (G_{o1}^*) were of interest (VanDongen et al. 1988). They identified the existence of several novel G-protein-gated, more precisely G_o-gated K^+, channels that are distinct from G_i-gated K^+ channels. Thus, application of purified bovine brain G_{o1}^* to the cytoplasmic aspect of inside-out membrane patches of cultured hippocampal pyramidal cells resulted in the appearance of three new types

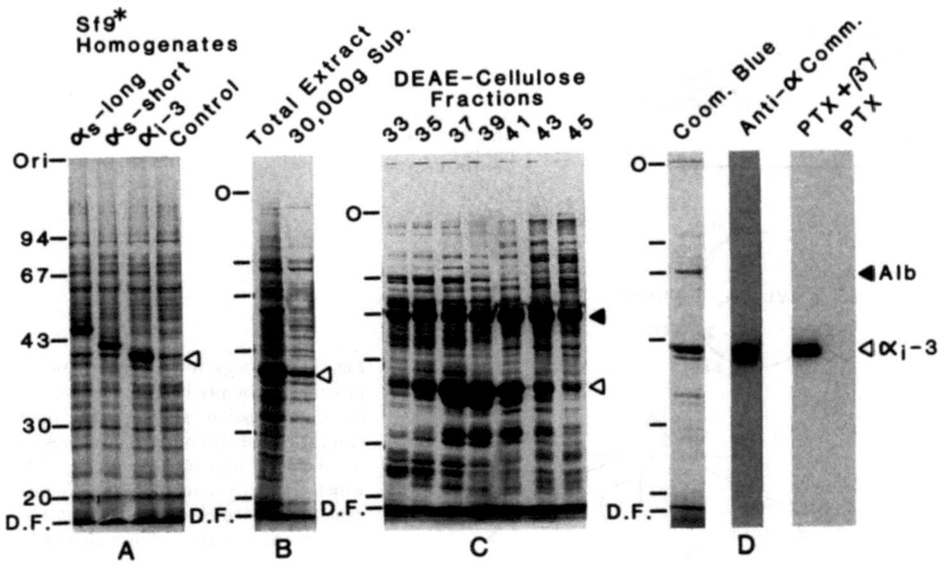

*Recombinant baculovirus (Autographa californica) in Spodoptera frugiperda (Sf) 9 cells

Fig. 14.A Baculoviral polyhedrin promoter directs the synthesis of recombinant α-subunits in Sf9 cells. **B** solubilization of recombinant (*bv*) α_{i3} from Sf9 cells. **C** partial purification of *bv* α_{i3} with DEAE-chromatography **D** western blot of partially purified *bv* α_{i3} and $\beta\gamma$ dimer dependent ADP-ribosylation of *bv* α_{i3} by pertussis toxin

of single channel K^+ currents consistent with the existence of the three nonrectifying K^+ channels having sizes of 13, 40, and 55 pS, respectively, plus an inwardly rectifying K^+ channel with a slope conductance of 40 pS. No such channel activities were observed with hRBC G_i^*-3 or hRBC α_i^*-3. G_{o2}^* or α_{o2}^* have not been tested in this system as yet. The effect of increasing concentrations of bovine brain G_{o1}^* on the 55 pS-type K^+ channel is shown in Fig. 16A. Note that in contrast to earlier observations with the same preparation of G_o^* added to guinea pig atrial membrane patches showing only marginal effects of GTPγS-activated G_{o1} at 2 nM (Yatani et al. 1987b), the hippocampal K^+ channel is highly sensitive to G_{o1}^*. Significant activation was obtained at 1 pM and half-maximal effects were obtained at about 10 pM. The identity of the active G protein in the G_{o1} preparations used was confirmed with recombinant GTPγS-activated α_{o1} (Fig. 16B). The G_o-gated channels were stimulated in the absence of Ca^{2+} or ATP, in the presence or absence of AMP-P(NH)P, added routinely to inhibit ATP-sensitive 70 pS K^+ channels. Further, EGTA did not interfere with the actions of G_{o1} or *recombinant* α_{o1}. Thus, in hippocampal pyramidal cells of the rat, G_{o1} is a G_K, and the K^+ channels gated by it are several and differ from those present in atrial cells in various aspects, including G protein specificity. Figure 17 schematizes these findings and raises the question which, if any, of the G proteins that gate K^+ channels regulate the other known PTX sensitive effector systems.

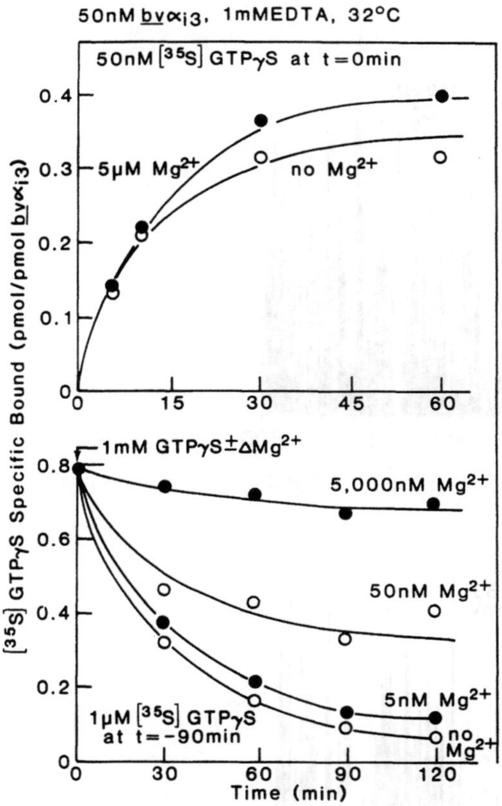

Fig. 15. Binding of GTPγS to recombinant $bv\ \alpha_{i3}$ is influenced by nanomolar concentrations of Mg^{2+}, thus resembling in this respect the GTPase activity of native G_i protein (Sunyer et al. 1984) and the GTPγS binding properties of the α-subunits of G_o (Higashijima et al. 1987)

4 βγ-Dimers Inhibit K⁺ Channel Gating by G Protein

Logothetis et al. (1987, 1988) reported twice that βγ-dimers stimulate atrial K^+ channel activity. Their finding is not reproduced in our hands. Quite the contrary, when we add βγ-dimers to inside-out membrane patches in which K^+ channels have been stimulated either by GTP only (baseline activity) or by carbachol plus GTP (agonist-stimulated activity), we find consistently inhibition of activity (Fig. 18). On their own, i.e., when added to silent patches in the absence of GTP, βγ-dimers have no effect under our assay conditions (Fig. 18; Okabe et al. 1989).

5 Current Views on How Signal Transduction by GTP and Receptors Comes About and Which Receptor Acts on Which G Protein to Regulate Which Effector System

Taken together the results discussed in the previous sections lead to several conclusions: (1) ionic channels are targets of direct regulation by G proteins as are adenylyl cyclase and the cGMP-specific phosphodiesterase; (2) α-subunits and not βγ-dimers

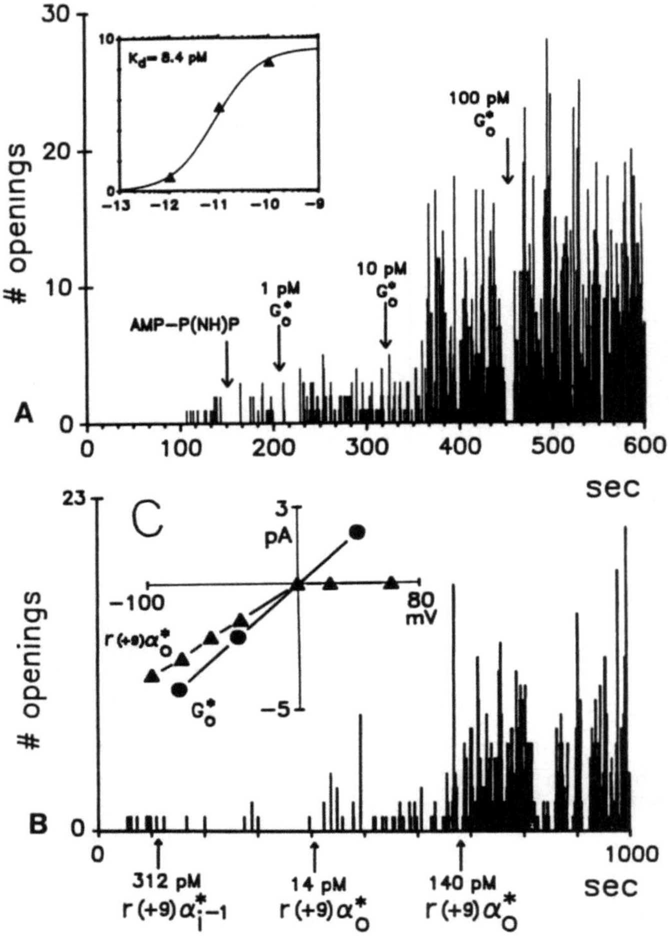

Fig. 16. Bovine brain G_o (**A**) and *recombinant* α_o, but not *recombinant* α_i-1 (**B**), gate single K^+ channel currents in hippocampal pyramidal cells of neonatal rats. **A** The effect of increasing concentrations of G_o^* on a 55 pS K^+ channel is shown by plotting the number of openings per 0.8 s as a function of time of continuous recording. *Inset* Openings per 0.8 s averaged for 1 min as a function of G_o^* concentration. **B** Lack of effects of 312 pM *recombinant* α_i^*-1, and stimulatory effect of increasing concentrations of *recombinant* α_o^* on a 40 pS K^+ channel, plotted as number of opening per 0.2 s as a function of the time of continuous recording shown. Single K^+ channel currents were recorded from excised inside-out membrane patches (Hamill et al. 1981) in the presence of 0.2 mM AMP-P(NH)P. Holding potential was –80 mV and both pipette and bath solutions contained 140 mM K-methanesulfonate, 1 mM EGTA, 1 mM $MgCl_2$, and 10 mM Hepes adjusted to pH 7.4 with Tris-base (Adapted from VanDongen et al. 1988)

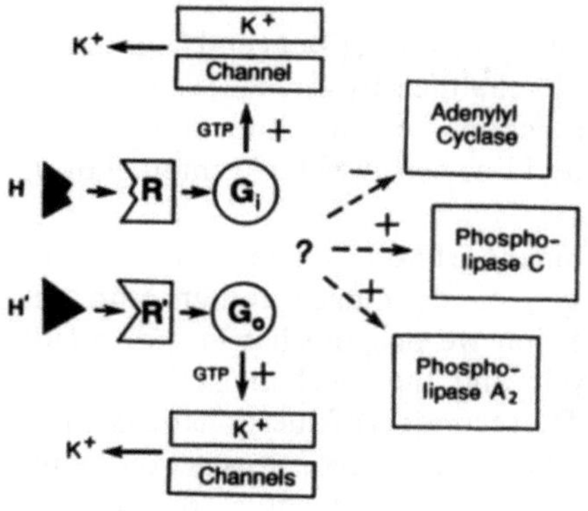

Fig. 17. The scheme depicts the existence of G_i-gated and G_o-gated K^+ channels and the fact that each of the G proteins may be pleotypic in its effects, i.e., able to regulate more than one effector function

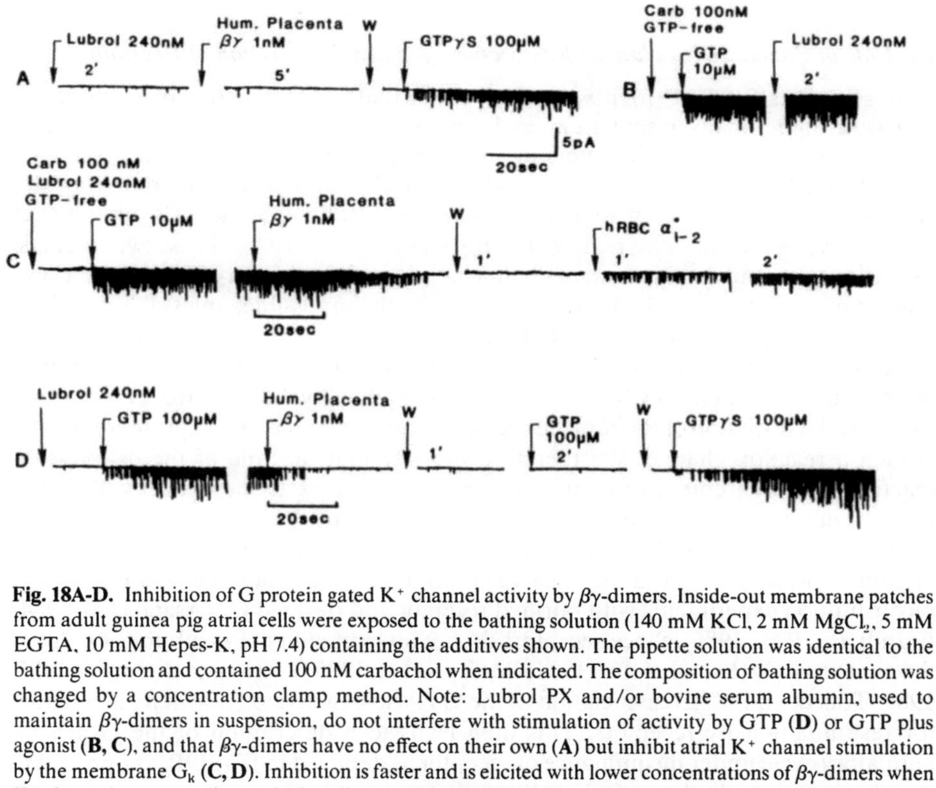

Fig. 18A-D. Inhibition of G protein gated K^+ channel activity by $\beta\gamma$-dimers. Inside-out membrane patches from adult guinea pig atrial cells were exposed to the bathing solution (140 mM KCl, 2 mM $MgCl_2$, 5 mM EGTA, 10 mM Hepes-K, pH 7.4) containing the additives shown. The pipette solution was identical to the bathing solution and contained 100 nM carbachol when indicated. The composition of bathing solution was changed by a concentration clamp method. Note: Lubrol PX and/or bovine serum albumin, used to maintain $\beta\gamma$-dimers in suspension, do not interfere with stimulation of activity by GTP (**D**) or GTP plus agonist (**B, C**), and that $\beta\gamma$-dimers have no effect on their own (**A**) but inhibit atrial K^+ channel stimulation by the membrane G_K (**C, D**). Inhibition is faster and is elicited with lower concentrations of $\beta\gamma$-dimers when K^+ channels are operating under baseline conditions (GTP only, experiment **D**: cessation of activity after 16 s) than when they are stimulated by agonist (carbachol plus GTP: experiment **C**: cessation of activity after 50 s). *Numbers* above records denote time elapsed in min between solution change and the beginning of the record shown

are the specificity determinants of signal transduction pathways; (3) several G proteins may have the same function, e.g., stimulation of K^+ channels by three G_i's; and (4) a single G protein may have more than one function, i.e., multifunctional (e.g., stimulation of adenylyl cyclase and the dihydropyridine-sensitive Ca^{2+} channel by a single G_s).

Two independent sets of questions emerge from these findings. The first deals with the subunit dissociation reaction, and asks what type of advantage it confers onto the system by its existence. The answer to this question we propose lies in a detailed analysis of the whole G-protein regulatory cycle and the mechanism by which receptors promote G-protein activation by GTP. The second set of questions deals with cross-talk between signal transduction pathways, i.e., whether receptors act on more than one G proteins, and if so which, whether G proteins interact with more than one receptor as well as with more than one effector, and if so, how frequently.

5.1 Role of Subunit Dissociation: Requirement for Catalytic Action of Receptors

In the second half of the 1970s it was demonstrated that receptors act catalytically rather than stoichiometrically to activate adenylyl cyclase (Tolkovsky and Levitzki 1978) and in 1980 it was found that receptors in addition to promoting GDP/GTP exchange (Cassel and Selinger 1978; Cassel et al. 1979) also participate in promoting the activation reaction proper of adenylyl cyclase by GTP and its analogs (Birnbaumer et al. 1980; Iyengar et al. 1980; reviewed in Birnbaumer et al. 1985). These two findings, catalytic action and stabilization of the nucleotide-activated form of adenylyl cyclase, now known to be in fact G_s, are thermodynamically impossible, unless the stabilized form of the G protein undergoes some type of additional spontaneous change that releases it from microscopic reversibility constraints, which happens in all enzymatically catalyzed reactions where the product is chemically different from the starting substrate. Dissociation of $\beta\gamma$ from the activated and stable receptor-G protein complex is such a reaction change. We therefore propose that the role of the dissociation reaction is a requirement without which receptors cannot act catalytically. Both the catalytic nature of receptor-mediated activation of G_s and the fact that receptor affects not only GDP release but also the rate at which inactive guanine nucleotide occupied G protein isomerizes from an inactive to an active state have been confirmed in reconstitution experiments with purified receptor and purified G_s (Asano et al. 1984; Brandt and Ross 1986). Since receptors do not interact with the α-subunits except in the context of $\beta\gamma$ (Kanaho et al. 1984; Watkins et al. 1985; Florio and Sternweis 1985, 1989; Tota et al. 1987), reassociation of the GTPase-deactivated α with $\beta\gamma$ is essential for restimulation by receptors. This then leads to a description of the G-protein regulatory cycle under the influence of receptor as depicted in Fig. 19B, as opposed to the alternative mode shown in Fig. 19A.

5.2 Specificities in Receptor-G Protein-Effector Interaction

Ever since the discoveries in the late 1960s that up to five different hormone receptors can activate in a single adenylyl cyclase system in an isolated membrane (Birnbaumer and Rodbell 1969), and, in the early and mid-1970s, that receptors can be transferred from one cell to another (Citri and Schramm 1980) and that there are no species and/or

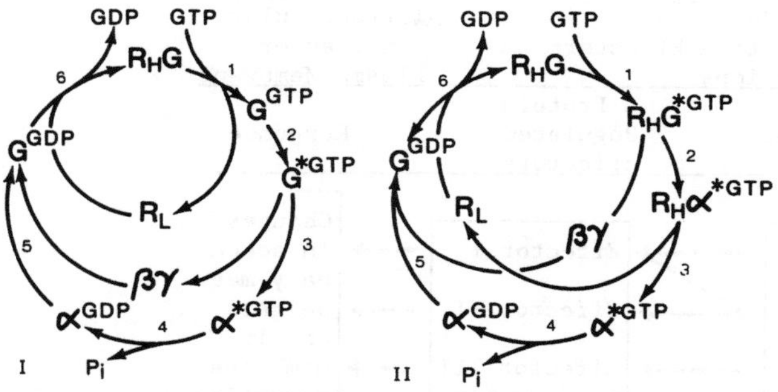

Fig. 19. Two models for catalysis of G-protein activation by hormone-receptor complex. In Model **I**, *Activation by Nucleotide Exchange,* the role of receptors is solely to promote nucleotide exchange. In Model **II**, *Activation by Subunit Dissociation,* the role of receptors is both to promote nucleotide exchange and to stabilize a GTP-dependent "activated" form of the G protein. In both models the G protein undergoes a cyclic dissociation-reassociation reaction and oscillates between GDP, nucleotide-free, and GTP states. The cycle is driven energetically forward by the capacity of the G protein to hydrolyze GTP (Cassel and Selinger 1978; Godchaux and Zimmerman 1979; Brandt and Ross 1986; Cerione et al. 1984). In both models the receptor has high affinity for agonist (R_H) when associated with the nucleotide-free trimeric $\alpha\beta\gamma$ form of the G protein, and has low affinity for the agonist (R_L) when it is free. In both models the receptor has higher affinity for the trimeric $\alpha\beta\gamma$ form of G than the G-GDP thus accounting for the finding that GDP and GDP analogs promote the R_H to R_L transition, and in both models $\beta\gamma$-dimers are required for the interaction of α with R. Thus, both models are identical in their reactions *4, 5,* and *6.* The models differ fundamentally in reaction *1*. Model **I** assumes reaction *1* to be symmetric to reaction *6,* with receptor merely allowing the GDP-GTP exchange to occur passively, and that reactions *2* and *3,* are thermodynamic consequences of the formation of G-GTP. Further, although the dissociation of G* to α* plus $\beta\gamma$ is indicated (reaction *3*) it is not an obligatory feature if only the catalytic nature of receptor action needs to be accounted for. Model **II**, on the other hand, assumes that no G* forms unless it is "aided" by receptor, and hence that receptor should have even higher affinity for the G*-GTP state than the nucleotide free state of *G.* As a consequence, this model is absolutely dependent on reaction *2* (subunit dissociation) to account for the catalytic nature of receptor action. Thermodynamic reasons do not allow R both to stabilize G* state and to dissociate from it. Reaction *3* of model **II** states further that the α*GTP loses its ability to stay associated with receptor and decomposes further into free-activated α*GTP plus free receptor, thus accounting for the fact that under "working" conditions (saturation by both GTP and hormone, and hence sustained regulation of effector) only a small proportion of receptors are found in their high affinity, G-protein associated state. Thus, in model **II**, the G-protein cycle is driven forward not only by the GTPase but also, and obligatorily so, by the subunit dissociation reaction. Since it has been shown experimentally (1) that receptors act catalytically and therefore need to dissociate from the G protein at one time or another (Tolkovsky and Levitzki 1978); and (2) that receptors accelerate the transition from inactive G-GTPγS to active G*-GTPγS transition and therefore must have higher intrinsic affinity for the activated than the inactive state (Birnbaumer et al. 1980; Iyengar et al. 1980; Iyengar and Birnbaumer 1982), it follows that model **II** does but model **I** does not account for the experimental findings

tissue specificity restrictions as to the source of G_s for reconstitution of a hormonally stimulable adenylyl cyclase system in *cyc*⁻ membranes (Ross et al. 1978; Kaslow et al. 1979), it has been clear that single G proteins are designed to interact with classes of receptors as opposed to single receptor subtypes. Figure 1 summarized the concepts derived from these types of experiments and presents some quantitative considerations as to the number of different receptors, G proteins, and effector functions one can think about at the present. Figure 20 presents an amplified version of the scheme shown in Fig. 1, which incorporates not only the findings presented here that one G protein can interact with more than one effector and that several G proteins may regulate a single

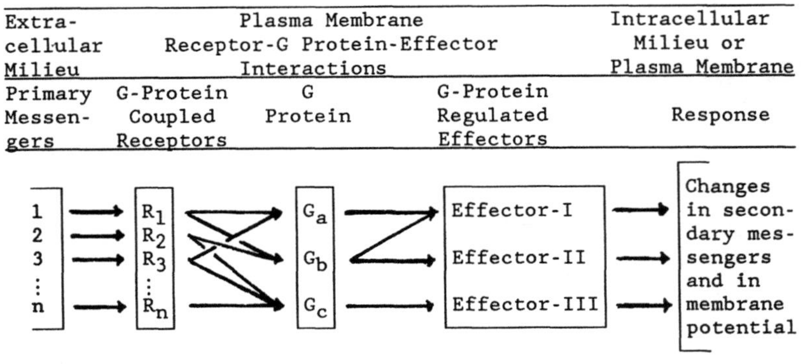

Fig. 20. Flow of information through G protein-dependent signal transduction systems of a cell with *n* receptors (R_1 through R_n), three G proteins (G_a through G_c), and three effector systems (*Effector-I* through *Effector-III*), each having separate and different selectivity for mutual interaction

effector, but also the findings by others that single receptors may affect more than one G protein as shown by Ashkenazi et al. (1987) for the M_2 acetylcholine receptor. The important notion that emerges from these considerations is that the wiring diagram describing signal transduction by G proteins needs to be determined individually and separately for each cell or tissue of interest. This includes the determination not only of which receptors are present but also of which G proteins and which effector functions are expressed.

Acknowledgments. Supported in part by United States Public Health Service research grants DK-19318, HD-09581, HL-31164 and HL-37044 to Lutz Birnbaumer, HL-39262 to Dr. Arthur M. Brown (Department of Physiology and Molecular Biophysics); a Welch Foundation grant (Q175) and the Baylor College of Medicine Diabetes and Endocrinology Research Center grant DK-27685 (Dr. James B. Field, Director).

References

Asano, T., Pedersen, S.E., Scott, C.W. & Ross, E.M. (1984) Reconstitution of catecholamine-stimulated binding of guanosine 5′-0-(3-thiotriphosphate) to the stimulatory GTP-binding protein of adenylate cyclase. Biochemistry 23:5460–5467

Ashkenazi, A., Winslow, J.W., Peralta, E.G., Peterson, G.L., Schimerlik, M.I., Capon, D.J. & Ramachandran, J. (1987) An M2 muscarinic receptor subtype coupled to both adenylyl cyclase and phosphoinositide turnover. Science 238:672–675

Beals, C.R., Wilson, C.B. & Perlmutter, R.M. (1987) A small multigene family encodes G_i signal-transduction proteins. Proc. Natl. Acad. Sci. USA 84:7886–7890

Birnbaumer, L. & Rodbell, M. (1969) Adenyl cyclase in fat cells. II. Hormone receptors. J. Biol. Chem. 244:3477–3482

Birnbaumer, L., Swartz, T.L., Abramowitz, J., Mintz, P.W. & Iyengar, R. (1980) Transient and steady state kinetics of the interaction of nucleotides with the adenylyl cyclase system from rat liver plasma membranes: interpretation in terms of a simple two-state model. J. Biol. Chem. 255:3542–3551

Birnbaumer, L., Hildebrandt, J.D., Codina, J., Mattera, R., Cerione, R.A., Sunyer, T., Rojas, F.J., Caron, M.G., Lefkowitz, R.J. & Iyengar, R. (1985) Structural basis of adenylate cyclase stimulation and inhibition by distinct guanine nucleotide regulatory proteins. In: Cohen, P. & Houslay, M.D. (eds) Molecular mechanisms of signal transduction. Elsevier/North Holland Biomedical Press, Amsterdam, pp. 131–182

Birnbaumer, L., Codina, J., Yatani, A., Mattera, R., Graf, R., Olate, J., Themmen, A.P.N., Liao, C.-F., Sanford, J., Okabe, K., Imoto, Y., Zhou, Z., Abramowitz, J., Suki, W.S., Hamm, H.E., Iyengar, R., Birnbaumer, M. & Brown, A.M. (1989) Molecular basis of regulation of ionic channels by G proteins. Rec. Prog. Hormone Res. (in press)

Brandt, D.R. & Ross, E.M. (1986) Catecholamine-stimulated GTPase cycle. Multiple sites of regulation by beta-adrenergic receptor and Mg^{2+} studied in reconstituted receptor-G_s vesicles. J. Biol. Chem. 261:1656–1664

Bray, P., Carter, A., Simons, C., Guo, V., Puckett, C., Kamholz, J., Spiegel, A. & Nirenberg, M. (1986) Human cDNA clones for four species of G *alpha*$_s$ signal transducing protein. Proc. Natl. Acad. Sci. USA 83:8893–8897

Bray, P., Carter, A., Guo, V., Puckett, C., Kamholz, J., Spiegel, A. & Nirenberg, M. (1987) Human cDNA clones for an α subunit of G_i signal-transduction protein. Proc. Natl. Acad. Sci. USA 84:5115–5119

Breitweiser, G.E. & Szabo, G. (1985) Uncoupling of cardiac muscarinic and beta-adrenergic receptors from ion channels by a guanine nucleotide analogue. Nature (London) 317:538–540

Brown, A.M. & Birnbaumer, L. (1988) Direct G protein gating of ion channels. Am. J. Physiol. 23:H401–H410

Cassel, D. & Selinger, Z. (1978) Mechanism of adenylate cyclase activation through the *beta*-adrenergic receptor: catecholamine-induced displacement of bound GDP by GTP. Proc. Natl. Acad. Sci. USA 75:4155–4159

Cassel, D., Eckstein, F., Lowe, M. & Selinger, Z. (1979) Determination of the turn-off reaction for the hormone-activated adenylate cyclase. J. Biol. Chem. 254:9835–9838

Cerbai, E., Klöckner & Isenberg, G. (1988) The α subunit of the GTP binding protein activates muscarinic potassium channels of the atrium. Science 240:1782–1783

Cerione, R.A., Codina, J., Benovic, J.L., Lefkowitz, R.J., Birnbaumer, L. & Caron, M.G. (1984) The mammalian beta$_2$-adrenergic receptor: reconstitution of the pure receptor with the pure stimulatory nucleotide binding protein (N_s) of the adenylate cyclase system. Biochemistry 23:4519–4525

Citri, Y. & Schramm, M. (1980) Resolution, reconstitution and kinetics of the primary action of hormone receptor. Nature (London) 287:297–300

Codina, J., Stengel, D., Woo, S.L.C. & Birnbaumer, L. (1986) Beta subunits of the human liver G_s/G_i signal transducing proteins and those of bovine retinal rod cell transducin are identical. FEBS Lett. 207:187–192

Codina, J., Yatani, A., Grenet, D., Brown, A.M. & Birnbaumer, L. (1987a) The *alpha* subunit of the GTP binding protein G_k opens atrial potassium channels. Science 236:442–445

Codina, J., Grenet, D., Yatani, A., Birnbaumer, L. & Brown, A.M. (1987b) Hormonal regulation of pituitary GH_3 cell K^+ channels by G_k is mediated by its *alpha* subunit. FEBS Lett. 216:104–106

Codina, J., Olate, J., Abramowitz, J., Mattera, R., Cook, R.G. & Birnbaumer, L. (1988) Alpha$_i$-3 cDNA encodes the alpha subunit of G_k, the stimulatory G protein of receptor-regulated K^+ channels. J. Biol. Chem. 263:6746–6750

Didsbury, J.R., & Snyderman, R. (1987) Molecular cloning of a new human G protein. Evidence for two $G_i\alpha$-like protein families. FEBS Lett. 219:259–263

Didsbury, J.R., Ho, Y.-S. & Snyderman, R. (1987) Human G_i protein alpha-subunit: deduction of amino acid structure from a cloned cDNA. FEBS Lett. 211:160–164

Dietzel, C. & Kurjan, J. (1987) The yeast SCG1 gene: a $G\alpha$-like protein implicated in the a- and α-factor response pathway. Cell 50:1001–1010

Ewald, D.A., Sternweis, P.C. & Miller, R.J. (1988) Guanine nucleotide-binding protein G_o-induced coupling of neuropeptide Y receptors to Ca^{2+} channels in sensory neurons. Proc. Natl. Acad. Sci. USA 85:3633–3637

Florio, V.A. & Sternweis, P.C. (1985) Reconstitution of resolved muscarinic cholinergic receptors with purified GTP-binding proteins. J. Biol. Chem. 260:3477–3483

Fong, H.K.W., Hurley, J.B., Hopkins, R.S., Miake-Lye, R., Johnson, M.S., Doolittle, R.F. & Simon, M.I. (1986) Repetitive segmental structure of the transducin beta subunit: homology with the CDC4 gene and identification of related mRNAs. Proc. Natl. Acad Sci USA 83:2162–2166

Fong, H.K.W., Amatruda, T.T., III, Birren, B.W.& Simon, M.I. (1987) Distinct forms of the β subunit of GTP-binding regulatory proteins identified by molecular cloning. Proc. Natl. Acad. Sci. USA 84:3792–3796

Fong, H.K., Yoshimoto, K.K., Eversole-Cire, P. & Simon, M.I. (1988) Identification of a GTP-binding protein α subunit that lacks an apparent ADP-ribosylation site for pertussis toxin. Proc. Natl. Acad. Sci 85:3066–3070

Gao, B., Gilman, A.G. & Robishaw, J.D. (1987) A second form of the β subunit of signal-transducing G proteins. Proc. Natl. Acad. Sci. USA 84:6122–6125

Gierschik, P., Codina, J., Simons, C., Birnbaumer, L. & Spiegel, A. (1985) Antisera against a guanine binding protein from retina cross-react with the *beta* subunit of the adenylyl cyclase associated guanine nucleotide binding proteins, N_s and N_i. Proc. Natl. Acad. Sci. USA 82:721–727

Godchaux, W., III & Zimmerman, W.F. (1979) Membrane-dependent guanine nucleotide binding and GTPase activities of soluble protein from bovine rod cell outer segments. J. Biol. Chem. 254:7874–7884

Graziano, M.P., Casey, P.J. & Gilman, A.G. (1987) Expression of cDNAs for G proteins in *Escherichia coli*. Two forms of $G_{s\alpha}$ stimulate adenylate cyclase. J. Biol. Chem. 262:11375–11381

Grunwald, G.B., Gierschik, P., Nirenberg, M. & Spiegel, A.M. (1986) Detection of α-transducin in retinal rods but not cones. Science 231:856–859

Halliday, K.R. (1983–84) Regional homology in GTP-binding proto-oncogene products and elongation factors. J. Cyclic Nucleotide & Protein Phosphorylat. Res. 9:435–448

Hamill, O.P., Marty, A., Neher, E., Sakmann, B. & Sigworth, F.J. (1981) Improved patch-clamp techniques for high resolution current recording from cells and cell-free membrane patches. Pflüger's Arch. 391:85–100

Harris, B.A., Robishaw, J.D., Mumby, S.M. & Gilman, A.G. (1985) Molecular cloning of complementary DNA for the alpha subunit of the G protein that stimulates adenylate cyclase. Science 229:1274–1277

Hescheler, J., Rosenthal, W., Trautwein, W. & Schultz, G. (1987) The GTP-binding protein, N_o, regulates neuronal calcium channels. Nature (London) 325:445–447

Hescheler, J., Rosenthal, W., Hinsch, K.-D., Wulfern, M., Trautwein, W. & Schultz, G. (1988) Angiotensin II-induced stimulation of voltage-dependent Ca^{2+} currents in an adrenal cortical cell line. EMBO J. 7:619–624

Higashijima, T., Ferguson, K.M., Sternweis, P.C., Smigel, M.D. & Gilman, A.G. (1987) Effects of Mg and the beta-gamma subunit complex on the interactions of guanine nucleotides with G proteins. J. Biol. Chem. 262:762–766

Holz, G.G., Rane, S.G. & Dunlap, K. (1986) GTP-binding proteins mediate transmitter inhibition of voltage-dependent calcium channels. Nature (London) 319:670–672

Hurley, J.B., Fong, H.K.W., Teplow, D.B., Dreyer, W.J. & Simon, M.I. (1984) Isolation and characterization of a cDNA clone for the gamma subunit of bovine retinal transducin. Proc. Natl. Acad. Sci. USA 81:6948–6952

Imoto, Y., Yatani, A., Reeves, J.P., Codina, J., Birnbaumer, L. & Brown, A.M. (1988) α Subunit of G_s directly activates cardiac calcium channels in lipid bilayers. Am. J. Physiol. 255:H722–H728

Itoh, H., Kozasa, T., Nagata, S., Nakamura, S., Katada, T., Ui, M., Iwai, S., Ohtsuka, E., Kawasaki, H., Suzuki, K. & Kaziro, Y. (1986) Molecular cloning and sequence determination of cDNAs for the alpha subunits of the guanine nucleotide-binding proteins G_s, G_i, and G_o from rat brain. Proc. Natl. Acad. Sci. USA 83:3776–3780

Itoh, H., Toyama, R., Kozasa, T., Tsukamoto, T., Matsuoka, M. & Kaziro, Y. (1988) Presence of three distinct molecular species of G_i protein α subunit. Structure of rat cDNAs and human genomic DNAs. J. Biol. Chem. 263:6656–6664

Iyengar, R. & Birnbaumer, L. (1982) Hormone receptors modulate the regulatory component of adenylyl cyclases by reducing its requirement for Mg ion and increasing its extent of activation by guanine nucleotides. Proc. Natl. Acad. Sci. USA 79:5179–5183

Iyengar, R., Abramowitz, J., Riser, M. & Birnbaumer, L. (1980) Hormone receptor-mediated stimulation of the rat liver plasma membrane adenylyl cyclase system: nucleotide effects and analysis in terms of a two-state model for the basic receptor-affected enzyme. J. Biol. Chem. 255:3558–3564

Jones, D.T. & Reed, R.R. (1987) Molecular cloning of five GTP-binding protein cDNA species from rat olfactory neuroepithelium. J. Biol. Chem. 262:14241–14249

Jones, D.T. & Reed, R.R. (1988) Olfactory signal transduction utilizes a novel GTP-binding protein. Cold Spring Harbor Symp. Quant. Biol. 53 (in press)

Jurnak, F. (1985) Structure of the GDP domain of EF-Tu and location of the amino acids homologous to *ras* oncogene proteins. Science 230:32–36

Kameyama, M., Hescheler, J., Hofmann, F. & Trautwein, W. (1986) Modulation of Ca current during the phosphorylation cycle in the guinea pig heart. Pflüger's Arch. 407:121–128

Kanaho, Y., Tsai, S.-C., Adamik, R., Hewlett, E.L., Moss, J. & Vaughan, M. (1984) Rhodopsin-enhanced GTPase activity of the inhibitory GTP-binding protein of adenylate cyclase. J. Biol. Chem. 259:7378–7381

Kaslow, H.R., Farfel, Z., Johnson, G.L. & Bourne, H.R. (1979) Adenylate cyclase assembled in vitro: cholera toxin substrates determine different patterns of regulation by isoproterenol and guanosine 5′-triphosphate. Mol. Pharmacol. 15:472–483

Kim, S., Ang, S.L., Bloch, D.B., Bloch, K.D., Kawahara, Y., Tolman, C., Lee, R., Seidman, J.G. & Neer, E.J. (1988) Identification of cDNA encoding an additional α subunit of a human GTP-binding protein: expression of three α_i subtypes in human tissues and cell lines. Proc. Natl. Acad. Sci. USA 85:4153–4157

Kirsch, G., Yatani, A., Codina, J., Birnbaumer, L. & Brown, A.M. (1988) The *alpha* subunit of G_k activates atrial K^+ channels of chick, rat and guinea pig. Am. J. Physiol. 254 (Heart Circular Physiol. 23):H1200–H1205

Kozasa, T., Itoh, H., Tsukamoto, T. & Kaziro, Y. (1988) Isolation and characterization of the human G_s α gene. Proc. Natl. Acad. Sci. USA 85:2081–2085

Kurjan, J. (1989) G proteins in the yeast *Saccharomyces cerevisiae*. In: Iyengar, R. & Birnbaumer, L. (eds) G proteins. Academic Press, New York London Orlando (in press)

Lerea, C.L., Somers, D.E., Hurley, J.B., Klock, I.B. & Bunt-Milan, A.H. (1986) Identification of specific transducin *alpha* subunits in retinal rod and cone photoreceptors. Science 234:77–80

Lewis, D.L., Weight, F.F. & Luini, A. (1986) A guanine nucleotide-binding protein mediates the inhibition of voltage-dependent calcium current by somatostatin in a pituitary cell line. Proc. Natl. Acad. Sci. USA 83:9035–9039

Light, D.B., Ausiello, D. & Stanton, B.A. (1989) A GTP binding protein, α_i^*-3, directly activates a sodium-conducting ion channel in a renal epithelium. Science (in press)

Lochrie, M.A. & Simon, M.I. (1988) G protein multiplicity in eukaryotic signal transduction systems. Biochemistry 17:4957–4965

Lochrie, M.A., Hurley, J.B. & Simon, M.I. (1985) Sequence of the alpha subunit of photoreceptor G protein: homologies between transducin, ras and elongation factors. Science 228:96–99

Logothetis, D.E., Kurachi, Y., Galper, J., Neer, E.J. & Clapham, D.E. (1987) The $\beta\gamma$ subunits of GTP-binding proteins activate the muscarinic K^+ channel in heart. Nature (London) 325:321–326

Logothetis, D.E., Kim, D., Northup, J.K., Neer, E.J. & Clapham, D.E. (1988) Specificity of action of guanine nucleotide-binding regulatory protein subunits on the cardiac muscarinic K^+ channel. Proc. Natl. Acad. Sci. USA 85:5814–5818

Luckov, V.A. & Summers, M.D. (1988) Trends in the development of baculovirus expression vectors. Bio/Technology 6:47–55

Masters, S.B., Stroud, R.M. & Bourne, H.R. (1986) Family of G protein *alpha* chains: Amphipatic analysis and predicted structure of functional domains. Protein Eng. 1:47–54

Matsuoka, M., Itoh, H., Kozasa, T. & Kaziro, Y. (1988) Sequence analysis of cDNa and genomic DNA for a putative pertussis toxin insensitive guanine nucleotide binding regulatory protein α subunit. Proc. Natl. Acad. Sci. USA 85:5384–5388

Mattera, R., Codina, J., Crozat, A., Kidd, V., Woo, S.L.C. & Birnbaumer, L. (1986) Identification by molecular cloning of two forms of the alpha subunit of the human liver stimulatory (G_s) regulatory component of adenylyl cyclase. FEBS Lett. 206:36–42

Mattera, R., Yatani, A., Kirsch, G.E., Graf, R., Olate, J., Codina, J., Brown, A.M. & Birnbaumer, L. (1989a) Recombinant α_i-3 subunit of G protein activates G_k-gated K^+ channels. J. Biol. Chem. 264:465–471

Mattera, R., Graziano, M.P., Yatani, A., Zhou, Z., Graf, R., Codina, J., Birnbaumer, L., Gilman, A.G. & Brown, A.M. (1989b) Individual splice variants of the α subunit of the G protein G_s activate both adenylyl cyclase and Ca^{2+} channels. Science 243:804–807

Medynski, D.C., Sullivan, K., Smith, D., Van Dop, C., Chang, F.-H., Fung, B.K.-K., Seeburg, P.H. & Bourne, H.R. (1985) Amino acid sequence of the alpha subunit of transducin deduced from the cDNA sequence. Proc. Natl. Acad. Sci. USA 82:4311–4315

Miyajima, I., Nakafuku, M., Nakayama, N., Brenner, C., Miyajima, A., Kalbuchi, K., Arai, K.-I., Kaziro, Y. & Matsumoto, K. (1987) GPA1, a haploid-specific essential gene, encodes a yeast homolog of mammalian G protein which may be involved in mating factor signal transduction. Cell 50:1011–1019

Nakafuku, M., Itoh, H., Nakamura, S. & Kaziro, Y. (1987) Occurrence in *Saccharomyces cerevisiae* of a gene homologous to the cDNA coding for the α subunit of mammalian G proteins. Proc. Natl. Acad. Sci. USA 84:2140–2144

Nakafuku, M., Obara, T., Kaibuchi, K., Miyajima, I., Itoh, H., Nakamura, S., Arai, K.-I., Matsumoto, K. & Kaziro, Y. (1988) Isolation of a second yeast *Saccharomyces cerevisiae* gene (*GPA2*) coding for guanine nucleotide-binding regulatory protein: studies on its structure and possible functions. Proc. Natl. Acad. Sci. USA 85:1374–1378

Nargeot, J., Nerbonne, J.M., Engels, J. & Lester, H.A. (1983) Time course of the increase in the myocardial slow inward current after a photochemically generated concentration jump of intracellular cAMP. Proc. Natl. Acad. Sci. USA 80:2395–2399

Nukada, T., Tanabe, T., Takahashi, H., Noda, M., Haga, K., Haga, T., Ichiyama, A., Kangawa, K., Hiranaga, M., Matsuo, H. & Numa, S. (1986a) Primary structure of the alpha-subunit of bovine adenylate cyclase inhibiting G-protein deduced from the cDNA sequence. FEBS Lett. 197:305–310

Nukada, T., Tanabe, T., Takahashi, H., Noda, M., Hirose, T., Inayama, S. & Numa, S. (1986b) Primary structure of the alpha-subunit of bovine adenylate cyclase-stimulating G-protein deduced from the cDNA sequence. FEBS Lett. 195:220–224

Okabe, K., Yatani, A., Evans, T., Ho, K.-J., Codina, J., Birnbaumer, L., Brown, A.M. (1989) $\beta\gamma$ dimers of G proteins: inhibition of receptor-mediated activation of atrial K^+ channels and lack of specificity for interaction with α subunits. (submitted)

Parent, L. & Coronado, R. (1988) K(ATP) channel is regulated by a G protein-dependent process in reconstituted T-tubule membranes from rabbit skeletal muscle. Biophys. J. Abstr.

Pfaffinger, P.J., Martin, J.M., Hunter, D.D., Nathanson, N.M. & Hille, B. (1985) GTP-binding proteins couple cardiac muscarinic receptors to a K channel. Nature (London) 317:536–538

Provost, N.M., Somers, D.E. & Hurley, J.B. (1988) A *Drosophila melanogaster* G protein α subunit gene is expressed primarily in embryos and pupae. J. Biol. Chem. 263:12070–12076

Ribalet, B., Ciani, S. & Eddlestone, G.T. (1989) Modulation of ATP-sensitive K channels in RINm5F cells by phosphorylation and G proteins. Biophys. J. Abstr.

Robishaw, J.D., Smigel, M.D. & Gilman, A.G. (1986a) Molecular basis for two forms of the G protein that stimulates adenylate cyclase. J. Biol. Chem. 261:9587–9590

Robishaw, J.D., Russell, D.W., Harris, B.A., Smigel, M.D. & Gilman, A.G. (1986b) Deduced primary structure of the alpha subunit of the GTP-binding stimulatory protein of adenylate cyclase. Proc. Natl. Acad. Sci. USA 83:1251–1255

Rosenthal, W., Hescheler, J., Hinsch, K.-D., Spicher, K., Trautwein, W. & Schultz, G. (1988) Cyclic AMP-independent, dual regulation of voltage-dependent Ca^{2+} currents by LHRH and somatostatin in a pituitary cell line. EMBO J. 7:1627–1633

Ross, E.M., Howlett, A.C., Ferguson, K.M. & Gilman, A.G. (1978) Reconstitution of hormone-sensitive adenylate cyclase activity with resolved components of the enzyme. J. Biol. Chem. 253:6401–6412

Scherer, N.M., Toro, M.-J., Entman, M.L. & Birnbaumer, L. (1987) G protein distribution in canine cardiac sarcoplasmic reticulum and sarcolemma. Comparison to rabbit skeletal membranes and brain and erythrocyte G proteins. Arch. Biochem. Biophys. 259:431–440

Schubert, B., VanDongen, A.M.J., Kirsch, G.E. & Brown, A.M. (1989) Modulation of cardiac Na channels by beta-adreno receptors and the G protein G_s. Biophysical J. 55:229A

Scott, R.H. & Dolphin, A.C. (1986) Regulation of calcium currents by GTP analogue: potentiation of (—)-baclofen-mediated inhibition. Neurosci. Lett. 69:59–64

Soejima, M. & Noma, A. (1984) Mode of regulation of the ACh-sensitive K-channel by the muscarinic receptor in rabbit atrial cells. Pflüger's Arch. 400:424–431

Sugimoto, K., Nukada, T., Tanabe, T., Takahashi, H., Noda, M., Minamino, N., Kangawa, K., Matsuo, H., Hirose, T., Inayama, S. & Numa, S. (1985) Primary structure of the beta-subunit of bovine transducin deduced from the cDNA sequence. FEBS Lett. 191:235–240

Suki, W., Abramowitz, J., Mattera, R., Codina, J. & Birnbaumer, L. (1987) The human genome encodes at least three non-allelic G proteins with $alpha_i$-type subunits. FEBS Lett. 220:187–192

Sullivan, K.A., Liao, Y.-C., Alborzi, A., Beiderman, B., Chang, F.-H., Masters, S.B., Levinson, A.D. & Bourne, H.R. (1986) Inhibitory and stimulatory G proteins of adenylate cyclase: cDNA and amino acid sequences of the alpha chains. Proc. Natl. Acad. Sci. USA 83:6687–6691

Sunyer, T., Codina, J. & Birnbaumer, L. (1984) GTPase properties of N_i, the inhibitory regulatory component of adenylyl cyclase. J. Biol. Chem. 259:15447–15451

Tabor, S. (1987) Dissection of the bacteriophage T7 DNA replication by the overproduction of its essential genetic elements. Doct. Thesis, Harvard Univ., School of Medicine

Tabor, S. & Richardson, C.C. (1985) A bacteriophage T7RNA polymerase/promote system for controlled exclusive expression of specific genes. Proc. Natl. Acad. Sci. USA 82:1074–1078

Tabor, S., Huber, H.E. & Richardson, C.C. (1987) *Escherichia coli* thioredoxin confers processivity on the DNA polymerase activity of the gene 5 protein of bacteriophage T7. J. Biol. Chem. 262:16212–16223

Tanabe, T., Nukada, T., Nishikawa, Y., Sugimoto, K., Suzuki, H., Takahashi, H., Noda, M., Haga, T., Ichiyama, A., Kangawa, K., Minamino, N., Matsuo, H. & Numa, S. (1985) Primary structure of the alpha-subunit of transducin and its relationship to *ras* proteins. Nature (London) 315:242–245

Tolkovsky, A.M. & Levitzki, A. (1978) Mode of coupling between the β-adrenergic receptor and adenylate cyclase in turkey erythrocytes. Biochemistry 17:3795–3810

Tota, M.R., Kahler, K.R. & Schimerlik, M.I. (1987) Reconstitution of the purified porcine atrial muscarinic acetylcholine receptor with purified porcine atrial inhibitory guanine nucleotide binding protein. Biochemistry 26:8175–8182

VanDongen, A., Codina, J., Olate, J., Mattera, R., Joho, R., Birnbaumer, L. & Brown, A.M. (1988) Newly identified brain potassium channels gated by the guanine nucleotide binding (G) protein G_o. Science 242:1433–1437

Van Dop, C., Tsubokawa, M., Bourne, H.R. & Ramachandran, J. (1984) Amino acid sequence of retinal transducin at the site ADP-ribosylated by cholera toxin. J. Biol. Chem. 259:696–699

VanMeurs, K.P., Angus, W., Lavu, S., Kung, H.F., Czarnecki, S.K., Moss, J. & Vaughan, M. (1987) Deduced amino acid sequence of bovine retinal $G_{o\alpha}$: similarities to other guanine nucleotide-binding proteins. Proc. Natl. Acad. Sci. USA 84:3107–3111

Van Veen, T., Oestholm, T., Gierschick, P., Spiegel, A.M., Somers, R., Korf, H.W. & Klein, D.C. (1986) α-Transducin immunoreactivity in retinae and sensory pineal organs of adult vertebrates. Proc. Natl. Acad. Sci. USA 83:912–916

Watkins, P.A., Burns, D.L., Kanaho, Y., Liu, T.-Y., Hewlett, E.L. & Moss, J. (1985) ADP-ribosylation of transducin by pertussis toxin. J. Biol. Chem. 260:13478–13482

West, R.E., Jr., Moss, J., Vaughan, M., Liu, T. & Liu, T.-Y. (1985) Pertussis toxin-catalyzed ADP-ribosylation of transducin. Cystein 347 is the ADP-ribose acceptor site. J. Biol. Chem. 260:14428–14430

Whiteway, M., Hougan, L., Dignard, D., Thomas, D.Y., Bell, L., Saari, G.C., Grant, F.J., O'Hara, P. & MacKay, V.L. (1989) The STE4 and STE18 genes of yeast encode potential β and γ subunits of the mating factor receptor-coupled G protein. Cell 56:467–477

Yarfitz, S., Provost, N.M. & Hurley, J.B. (1988) Cloning of a *Drosophila melanogaster* guanine nucleotide regulatory protein β-subunit gene and characterization of its expression during development. Proc. Natl. Acad. Sci. USA 85:7134–7138

Yatani, A., Codina, J., Brown, A.M. & Birnbaumer, L. (1987a) Direct activation of mammalian atrial muscarinic potassium channels by GTP regulatory protein G_k. Science 235:207–211

Yatani, A., Codina, J., Sekura, R.D., Birnbaumer, L. & Brown, A.M. (1987b) Reconstitution of somatostatin and muscarinic receptor mediated stimulation of K^+ channels by isolated G_k protein in clonal rat anterior pituitary cell membranes. Mol. Endocrinol. 1:283–289

Yatani, A., Codina, J., Imoto, Y., Reeves, J.P., Birnbaumer, L. & Brown, A.M. (1987c) A G protein directly regulates mammalian cardiac calcium channels. Science 238:1288–1292

Yatani, A., Imoto, Y., Codina, J., Hamilton, S.L., Brown, A.M. & Birnbaumer, L. (1988a) The stimulatory G protein of adenylyl cyclase, G_s, directly stimulates dihydropyridine-sensitive skeletal muscle Ca^{2+} channels. Evidence for direct regulation independent of phosphorylation by cAMP-dependent protein kinase. J. Biol. Chem. 263:9887–9895

Yatani, A., Hamm, H., Codina, J., Mazzoni, M.R., Birnbaumer, L. & Brown, A.M. (1988b) A monoclonal antibody to the α subunit of G_k blocks muscarinic activation of atrial K^+ channels. Science 241:828–831

Yatani, A., Mattera, R., Codina, J., Graf, R., Okabe, K., Padrell, E., Iyengar, R., Brown, A.M. & Birnbaumer, L. (1988c) The G protein-gated atrial K^+ channel is stimulated by three distinct $G_i\alpha$-subunits. Nature (London) 336:680–682

Yatsunami, K. & Khorana, H.G. (1985) GTPase of bovine rod outer segments: the amino acid sequence of the alpha subunit as derived from the cDNA sequence. Proc. Natl. Acad. Sci USA 82:4316–4320

Yatsunami, K., Pandya, B.V., Oprian, D.D. & Khorana, H.G. (1985) cDNA-derived amino acid sequence of the gamma subunit of GTPase from bovine rod outer segments. Proc. Natl. Acad. Sci. USA 82:1936–1940

The Guanylate Cyclase Family

D.L. GARBERS[1]

1 Introduction

Following the discovery of adenylate cyclase (Sutherland et al. 1962), the enzyme guanylate cyclase was identified (Hardman and Sutherland 1969; Schultz et al. 1969; White and Aurbach 1969). By the time of its discovery, the "second messenger hypothesis" was well in place and it was initially felt that the mechanisms by which guanylate cyclase were regulated might fit the general model being developed for hormonal regulation of adenylate cyclase (Robison et al. 1967). However, that guanylate cyclase activity distributed to various cellular compartments, including the cytoplasm, and that for many years reproducible activation of guanylate cyclase in broken cell preparations was observed only with nitrovasodilators and similar agents, which interacted with the cytoplasmic form of the enzyme (Garbers 1989a), raised doubts about applications of the adenylate cyclase model to guanylate cyclase.

In the late 1970s and early 1980s, peptides obtained from egg-conditioned media (Hansbrough and Garbers 1981a; Suzuki et al. 1981) and from cardiac atria (deBold et al. 1981) were shown to cause elevations of cyclic GMP in various cells (Kopf and Garbers 1979; Hansbrough and Garbers 1981a; Hamet et al. 1984; Waldman et al. 1984). In both cases, the peptides were subsequently shown to alter guanylate cyclase activity in broken cell preparations (Waldman et al. 1984; Ramarao and Garbers 1985; Bentley et al. 1986b), and in both cases the enzyme activated was shown to be a form associated with the plasma membrane.

A major question became: "How does the binding of these peptides to putative cell surface receptors stimulate guanylate cyclase activity?" This chapter will address, in part, the answer to this question.

2 Results and Discussion

2.1 *Peptide Structures*

One of the primary reasons that activation of the membrane form of guanylate cyclase was not observed in broken cell preparations in early studies using many different known hormones was that direct activators of the enzyme had not yet been discovered.

[1]Howard Hughes Medical Institute and Departments of Pharmacology and Molecular Physiology and Biophysics, Vanderbilt Medical Center, Nashville, TN 37232–0295, USA

40. Colloquium Mosbach 1989
Molecular Mechanisms of Hormone Action

Table 1. Structures of some peptides known to stimulate guanylate cyclase activity[a]

Name	Structure	Source
Speract	GFDLNGGGVG	ECM
Resact	CVTGAPGCVGGGRL	"
Alloresact	KLCPGGNCV	"
Mosact	DSDSAQNLIG	"
ANP	SLRRSSCFGGRMDRIGAQSGLGCNSRFY	Rat heart
BNP	DSGCFGRRLDRIGSLSGLGCNVLRRY	Rat brain
STa	NTFYCCELCCNPACAGCY	*E. coli*
STa	NSSNYCCELCCNPACTGCY	*E. coli*

[a] The abbreviations used are single letters for the amino acids, ECM for egg-conditioned media, ANP for atrial natriuretic peptide, BNP for brain natriuretic peptide, and STa for heat-stable enterotoxins from *E. coli*.

The first peptide isolated from the egg-conditioned media of sea urchins was speract (Gly-Phe-Asp-Leu-Asn-Gly-Gly-Gly-Val-Gly), a decapeptide capable of activating sperm motility and respiration (Hansbrough and Garbers, 1981a,b; Suzuki et al. 1981). Within a few years, other peptides were isolated (Table 1), but these did not necessarily cross-react with spermatozoa from other species. Despite the species specificity of the egg peptides, the biochemical responses of the sperm cell to a given peptide remained essentially the same (Garbers 1988). Elevations of intracellular pH and Ca^{2+} (Repaske and Garbers 1983; Schackmann and Chock 1986), K^+ efflux (Lee and Garbers 1986), and transient increases in cyclic AMP and cyclic GMP (Hansbrough and Garbers 1981a) include some of the early changes observed in sperm cells after peptide addition. One of the earliest responses of spermatozoa to the peptides is the transient elevation of cyclic GMP (Hansbrough and Garbers 1981a).

Since the biochemical responses are similar or the same between the species, it has been suggested that the sperm surface receptors of different species retain conserved regions within the transmembrane or intracellular regions, but that extracellular regions vary to accomodate different ligands across the species (Garbers 1989b).

At about the same time the egg peptides were discovered, peptides isolated from cardiac tissue called atrial natriuretic peptides (deBold et al. 1981), were discovered. As can be seen from Table 1, these peptides bear no apparent resemblance to the egg peptides, but like the egg peptides, they too cause elevations of cyclic GMP in various target tissues (Hamet et al. 1984; Waldman et al. 1984).

Included in Table 1, but not discussed further, are the structures of heat-stable enterotoxins from *Escherchia coli,* which also stimulate a particulate guanylate cyclase activity (Field et al. 1978).

2.2 *Preliminary Identification of the Receptors*

Cross-linking reagents or photoaffinity probes were first used in attempts to identify the apparent receptors for the egg-derived or the atrial natriuretic peptides. For spermatozoa, two different cross-linked proteins were identified, one with an apparent molecular weight (M_r) of 77 000 (Dangott and Garbers 1984) and one with an M_r 160 000 (Shimomura et al. 1986). The protein cross-linked depended on the peptide, and therefore the species. The large M_r cross-linked protein was identified as guanylate cyclase based on comigration of the cross-linked protein and homogeneous guanylate cyclase, the precipitation of the cross-linked protein by antibody produced to guanylate cyclase, and a shift in the M_r of both guanylate cyclase and the cross-linked protein by the addition of NH_4 Cl to intact cells (Shimomura et al. 1986). The M_r 77 000 protein has not been identified as any previously known protein, and its sequence does not resemble that of guanylate cyclase (Dangott et al. 1989).

With atrial natriuretic peptides, two apparent receptors also were initially identified, one with an M_r 130 000 and one with an M_r 66 000 (Yip et al. 1985; Schenk et al. 1985; Vandlen et al. 1985; Leitman et al. 1986). There is also now evidence for a third receptor of M_r 180 000 (Paul et al. 1987). It has been suggested that the M_r 130 000 and 180 000 receptors are coupled to guanylate cyclase activation, and that the 66 000 M_r receptor is a "silent receptor" (Maack et al. 1987), but there is not full agreement on these conclusions (Ishido et al. 1989).

2.3 Sea Urchin Sperm Guanylate Cyclase

Gray et al. (1976) first described high activities of guanylate cyclase in invertebrate spermatozoa. Subsequent studies then demonstrated that the sperm enzyme resembled the particulate form of guanylate cyclase from various mammalian tissues with respect to both kinetic and immunologic characteristics (Garbers et al. 1974; Garbers 1978). The membrane form of the enzyme, first purified to apparent homogeneity from spermatozoa, became a model for the mammalian enzyme (Garbers 1976; Radany et al. 1983).

One mechanism by which the sea urchin sperm enzyme is desensitized to an egg peptide is now known. In 1983, Ward and Vacquier reported that egg jelly caused a decrease in the M_r of a plasma membrane protein of spermatozoa. The decrease in M_r was shown to be due to a loss of phosphate from serine residues (Ward and Vacquier 1983; Vacquier and Moy 1986). The M_r of the plasma membrane protein was similar to the M_r of homogeneous guanylate cyclase and experiments then confirmed that they were, in fact, the same protein (Suzuki et al. 1984; Ward et al. 1985). The substance in the egg jelly that caused the M_r shift was subsequently identified as the peptide, resact (Suzuki et al. 1984). Coincident with the change in M_r, guanylate cyclase activity markedly decreased (Ward et al. 1985; Ramarao and Garbers 1985). Subsequently, Bentley et al. (1986a) isolated plasma membrane vesicles from sea urchin spermatozoa that contained approximately 70% of the guanylate cyclase in the highly phosphorylated state. When resact was added to these membranes, guanylate cyclase was dephosphorylated, but prior to the loss of phosphate, activity markedly increased. The loss of phosphate coincided with a desensitization phenomenon in that guanylate cyclase activity declined as a result of dephosphorylation despite the continued presence of resact (Bentley et al. 1986a).

To determine the primary structure of the membrane form of guanylate cyclase and to provide a means to determine whether or not the enzyme could serve as a receptor, the mRNA for the protein was cloned. Amino acid sequences of proteolytic fragments of the purified sea urchin sperm guanylate cyclase were obtained, and oligonucleotide probes based on these sequences were then used to identify and isolate cDNA clones encoding the protein (Singh et al. 1988). The first mRNA to be cloned was that from *Arbacia punctulata* testis; it contained a single putative transmembrane domain that divides the enzyme roughly in one-half (Singh et al. 1988). Of notable interest was the presence of an intracellular domain homologous to the protein kinase family.

The *A. punctulata* cDNA clone also was used to isolate cDNA clones from another sea urchin (*Strongylocentrotus purpuratus*) species. The predicted amino acid sequences were very similar between the species, but the *S. purpuratus* sequence was extended by 136 amino acids (Thorpe and Garbers 1989).

When the *A. punctulata* or the *S. purpuratus* cDNA clones were expressed in cultured mammalian cells, there was no detectable expression of either guanylate cyclase or peptide binding activity.

2.4 Mammalian Guanylate Cyclase

Only the plasma membrane form of the mammalian enzyme will be discussed. In general, the activity of this form of the enzyme in various mammalian tissues is very low. Purification, then, has been difficult and only within the last few years have reasonably high specific activities of the enzyme been obtained after various purification procedures. In four different laboratories, enzyme activity has copurified with a protein that binds atrial natriuretic peptide (Kuno et al. 1986; Paul et al. 1987; Takayanagi et al. 1987; Meloche et al. 1988). The relatively low amounts of recovered protein, however, precluded definitive arguments as to whether or not the enzyme and ANP receptor were the same protein.

To determine whether or not guanylate cyclase could serve as the ANP receptor, the cDNA clone from the sea urchin was used as a probe to isolate similar sequences in the mammal. It was felt that if guanylate cyclase was a cell surface receptor that the catalytic domain would remain highly conserved. This, in fact, appears to be the case.

The sea urchin probe was used to initially identify partial length cDNA clones from the human (Lowe et al. 1989) and these clones were then used to isolate a full-length cDNA clone from a rat brain cDNA library (Chinkers et al. 1989). The predicted general topology of the rat brain enzyme was similar to that of the sea urchin guanylate cyclase, and the protein kinase-line domain remained conserved.

The binding of [^{125}I]-ANP to intact cells transfected with the rat brain cDNA increased dramatically (Chinkers et al. 1989). The binding characteristics were those expected for the high but not the low M_r receptor for ANP. Moreover, cross-linking studies using the intact, transfected cells demonstrated the appearance of a major, radiolabeled band of M_r 130 000 on Na•dodecyl•SO_4 polyacrylamide gels after electrophoresis. These data provided definitive evidence that guanylate cyclase is the high M_r ANP receptor.

In detergent extracts of the transfected cells, ANP stimulated guanylate cyclase activity approximately three fold, which is about the same as seen with detergent extracts of various mammalian cells. Studies also were conducted to see if ANP could elevate cyclic GMP in intact, transfected COS-7 cells. In the intact transfected cells, cyclic GMP was approximately the same as in control cells, suggesting that transfected cells are able to regulate cyclic GMP concentrations despite very high amounts of guanylate cyclase. Upon the addition of ANP, however, cyclic GMP concentrations increased over 100-fold within 1 min in the transfected cells but only by about 50% in the nontransfected cells, thereby demonstrating that the expressed protein could effectively transduce the ANP binding signal to the catalytic domain (Singh and Garbers, personal communication).

A human cDNA for guanylate cyclase also has yielded results similar to those described for the rat brain clone (Lowe et al. 1989).

2.5 *Members of the Guanylate Cyclase Family*

The membrane form of guanylate cyclase regulated by extracellular peptides and the cytoplasmic form regulated by nitrovasodilators contain at least one homologous domain (Thorpe and Garbers 1989; Chinkers et al. 1989; Lowe et al. 1989; Koesling et al. 1988; Nakane et al. 1988). The primary amino acid sequence of a second proposed subunit of the cytoplasmic form of the enzyme remains to be determined. Since the cytoplasmic subunit has not displayed catalytic activity (Nakane et al. 1988) and since one sea urchin enzyme (Singh et al. 1988) has contained only a small segment within this region, it has been suggested that this homologous domain is not the catalytic site. However, recent data from Krupinski et al. (1989) on the primary sequence of the bovine brain adenylate cyclase raised significant concerns with respect to this conclusion. Two putative cytoplasmic loops of adenylate cyclase are homologous to the guanylate cyclases in the carboxyl region where the cytoplasmic and membrane forms display a high degree of identity. This suggested that this is either a highly conserved regulatory element or a part of the catalytic site. Since there is also some conservation of the same primary sequence within the putative catalytic domain of a yeast adenylate cyclase (Chinkers et al. 1989), it seemed likely that this conserved element, in fact, would represent a catalytic site. Chinkers and Garbers (personal communication) have now, in fact, localized the catalytic site to the carboxyl region by the use of deletion mutants. The protein kinase-like domain appears to be regulatory in nature.

A low molecular weight receptor for ANP that has been termed the silent or clearance receptor has been cloned (Fuller et al. 1988). The receptor appears to contain only 37 amino acids within the cytoplasm, but within the extracellular domain, almost 34% identity exists at the primary amino acid sequence level, between this receptor and the rat brain ANP/guanylate cyclase receptor (Fuller et al. 1988; Chinkers et al. 1989).

2.6 *A New But Old Paradigm for Signal/Transduction*

Robison et al. (1967) provided early models to explain the hormonal regulation of adenylate cyclase, and one such model provided the possibility that hormone would bind directly to the enzyme to increase the production of cyclic AMP. The guanylate cyclase receptor now appears to fit this model, although associated regulatory proteins required for transduction of the binding signal may yet be involved. This is the first example of a cell surface receptor, which also possesses enzyme activity that is capable of forming a low molecular weight second messenger. The number of guanylate cyclase variants with receptor activity remains to be determined, but it would not be surprising to find ligands other than ANP that interact with guanylate cyclase in the mammal, or alternatively, guanylate cyclase subtypes which bind variant structures of ANP with different affinities.

References

Bentley, J.K., Shimomura, H. & Garbers, D.L. (1986a) Retention of a functional resact receptor in isolated sperm plasma membranes. Cell 45:281–288

Bentley, J.K., Tubb, D.J. & Garbers, D.L. (1986b) Receptor-mediated activation of spermatozoan guanylate cyclase. J. Biol. Chem. 261:14859–14862

Chinkers, M., Garbers, D.L., Chang, M.S. Lowe, D.G., Chin, H., Goeddel, D.V. & Schulz, S. (1989) A membrane form of guanylate cyclase is an atrial natriuretic peptide receptor. Nature (London) 338:78–83

Dangott, L.J. & Garbers, D.L. (1984) Identification and partial characterization of the receptor for speract. J. Biol. Chem. 259:13712–13716

Dangott, L.J., Jordan, J.E., Bellet, R.A. & Garbers, D.L. (1989) Cloning of the mRNA for the protein that crosslinks to speract. Proc. Natl. Acad. Sci. USA 86:2128–2132

deBold, A.U., Bornstein, H.B., Veress, A.T. & Sonnenberg, H. (1981) A rapid and potent natriuretic response to intravenous injection of atrial myocardial extract in rats. Life Sci. 28:89–94

Field, M., Graf, L.H., Jr., Laird, W.J., & Smith, P.L. (1978) Heat-stable enterotoxin of *Escherichia coli:* in vitro effects on guanylate cyclase activity, cyclic GMP concentration, and ion transport in small intestine. Proc. Natl. Acad. Sci. USA 75:2800–2804

Fuller, F., Porter, J.G., Arfsten, A.E., Miller, J., Schilling, J.W., Scarborough, R.M., Lewicki, J.A. & Schenk, D.B. (1988) Atrial natriuretic peptide clearance receptor. Complete sequence and functional expression of cDNA clones. J. Biol. Chem. 263:9395–9401

Garbers, D.L. (1976) Sea urchin sperm guanylate cyclase: purification and loss of cooperativity. J. Biol. Chem. 251:4071–4077

Garbers, D.L. (1978) Sea urchin sperm guanylate cyclase antibody. J. Biol. Chem. 253:1898–1901

Garbers, D.L. (1988) Signal/transduction mechanisms of sea urchin spermatozoa. ISI Atlas of Science: Biochemistry 1:120–126

Garbers, D.L. (1989a) Guanylate cyclase, a cell surface receptor. J. Biol. Chem. 264:9103–9106

Garbers, D.L. (1989b) Molecular basis of fertilization. Annu. Rev. Biochem. 58:719–742

Garbers, D.L., Hardman, J.G. & Rudolph, F.G. (1974) Kinetic analysis of sea urchin sperm guanylate cyclase. Biochemistry 13:4166–4171

Gray, J.P., Drummond, G.I., Luk, D.W., Hardman, J.G. & Sutherland, E.W. (1976) Enzymes of cyclic nucleotide metabolism in invertebrate and vertebrate spermatozoa. Arch. Biochem. Biophys. 172:20–30

Hamet, P., Tremblay, J., Pang, S.C., Garcia, R., Thibault, G., Gutkowski, J., Cantin, M. & Genest, J. (1984) Effect of native and synthetic atrial natriuretic factor on cyclic GMP. Biochem. Biophys. Res. Commun. 123:515–527

Hansbrough, J.R. & Garbers, D.L. (1981a) Speract: purification and characterization of a peptide associated with eggs that activates spermatozoa. J. Biol. Chem. 256:1447–1452

Hansbrough, J.R. & Garbers, D.L. (1981b) Sodium-dependent activation of sea urchin spermatozoa by speract and monensin. J. Biol. Chem. 256:2235–2241

Hardman, J.G. & Sutherland, E.W. (1969) Guanyl cyclase: an enzyme catalyzing the formation of guanosine 3′,5′-monophosphate from guanosine triphosphate. J. Biol. Chem. 244:6363–6370

Ishido, M., Fujita, T., Shimonaka, M., Saheki, T., Ohuchi, S., Kume, T., Ishigaki, I. & Hirose, S. (1989) Inhibition of atrial natriuretic peptide-induced cyclic GMP accumulation in the bovine endothelial cells with antiatrial natriuretic peptide receptor antiserum. J. Biol. Chem. 264:641–645

Koesling, D., Herz, J., Gausepohl, H., Niroomand, F., Hinsch, K.-D., Mulsch, A., Bohme, E., Schultz, G. & Frank, R. (1988) The primary structure of the 70 kDa subunit of bovine soluble guanylate cyclase. FEBS Lett. 239:29-34

Kopf, G.S. & Garbers, D.L. (1979) A low molecular weight factor from sea urchin eggs elevates sperm cyclic nucleotide concentrations and respiration rates. J. Reprod. Fertil. 57:353–361

Krupinski, J., Coussen, F., Bakalyar, H.A., Tang, W.-J., Feinstein, P.G., Orth, K., Slaughter C., Reed, R.R. & Gilman, A.G. (1989) Adenylyl cyclase amino acid sequence: possible channel- or transporter-like structure Science 244:1558–1564

Kuno, T., Andresen, J.W., Kamisaki, Y., Waldman, S.A., Chang, L.Y., Saheki, S., Leitman, D.C., Nakane, M. & Murad, F. (1986) Co-purification of an atrial natriuretic factor receptor and particulate guanylate cyclase from rat lung. J. Biol. Chem. 261:5817–5823

Lee, H.C. & Garbers, D.L. (1986) Modulation of the voltage-sensitive Na^+/H^+ exchange in sea urchin spermatozoa through membrane potential changes induced by the egg peptide speract. J. Biol. Chem. 261:16026–16032

Leitman, D.C., Andresen, J.W., Kuno, T., Kamisaki, Y., Chang, J.-K. & Murad, F. (1986) Identification of multiple binding sites for atrial natriuretic factor by affinity cross-linking in cultural endothelial cells. J. Biol. Chem. 261:11650–11655

Lowe, D.G., Chang, M.S., Hellmis, R., Chen, E., Singh, S., Garbers, D.L. & Goeddel, D.V. (1989) Human atrial natriuretic peptide receptor defines a new paradigm for second messenger signal transduction. EMBO J. 8:1377–1384

Maack, T., Suzuki, M., Almeida, F.A., Nossenzweig, D., Scarborough, R.M., McEnroe, G.A. & Lewicki, J.A. (1987) Physiological role of silent receptors of atrial natriuretic factor. Science 238:675–678

Meloche, S., McNicoll, N., Liu, B., Ong, H. & DeLean, A. (1988) Atrial natriuretic factor R_1 receptor from bovine adrenal glomerulosa: purification, characterization, and modulation by amiloride. Biochemistry 27:8151–8158

Nakane, M., Saheki, S., Kuno, T., Ishii, K. & Murad, F. (1988) Molecular cloning of a cDNA coding for 70 kilodalton subunit of soluble guanylate cyclase from rat lung. Biochem. Biophys. Res. Commun. 157:1139–1147

Paul, A.K., Marala, R.B., Jaiswal, R.K. & Sharma, R.K. (1987) Coexistence of guanylate cyclase and atrial natriuretic factor receptor in a 180-kD protein. Science 235:1224–1226

Radany, E.W., Gerzer, R. & Garbers, D.L. (1983) Purification and characterization of particulate guanylate cyclase from sea urchin spermatozoa. J. Biol. Chem. 258:8346–8351

Ramarao, C.S. & Garbers, D.L. (1985) Receptor-mediated regulation of guanylate cyclase activity in spermatozoa. J. Biol. Chem. 260:8390–8396

Repaske, D.R. & Garbers, D.L. (1983) A hydrogen ion flux mediates stimulation of respiratory activity by speract in sea urchin spermatozoa. J. Biol. Chem. 258:6025–6029

Robison, G.A., Butcher, R.W. & Sutherland, E.W. (1967) Adenyl cyclase as an adrenergic receptor. Ann. N.Y. Acad. Sci. 139:703–723

Schackmann, R.W. & Chock, B.P. (1986) Alteration of intracellular $[Ca^{2+}]$ in sea urchin sperm by the egg peptide speract. Evidence that increased intracellular Ca^{2+} is coupled to Na^+ entry and increased intracellular pH. J. Biol. Chem. 261:8719–8728

Schenk, D.B., Phelps, M.N., Porter, J.G., Scarborough, R.M., McEnroe, G.A. & Lewicki, J.A. (1985) Identification of the receptor for atrial natriuretic factor on cultured vascular cells. J. Biol. Chem. 260:14887–14890

Schultz, G., Bohme, E. & Muske, K. (1969) Guanyl cyclase: Determination of enzyme activity. Life Sci. 8:1323–1332

Shimomura, H., Dangott, L.J. & Garbers, D.L. (1986) Covalent coupling of a resact analogue to guanylate cyclase. J. Biol. Chem. 261:15778–15782

Singh, S., Lowe, D.G., Thorpe, D.W., Rodriguez, H., Kuang, W.-J., Dangott, L.J., Chinkers, M., Goeddel, D.V. & Garbers, D.L. (1988) Membrane guanylate cyclase is a cell surface receptor with homology to protein kinases. Nature (London) 334:708–712

Sutherland, E.W., Rall, T.W. & Menon, T. (1962) Adenyl cyclase I. Distribution, preparation and properties. J. Biol. Chem. 237:1220–1227

Suzuki, N., Nomura, K., Ohtake, H. & Isaka, S. (1981) Purification and the primary structure of sperm-activating peptides from the jelly coat of sea urchin eggs. Biochem. Biophys. Res. Commun. 99:1238–1244

Suzuki, N., Shimomura, H., Radany, E.W., Ramarao, C.S., Ward, G.E., Bentley, J.K. & Garbers, D.L. (1984) A peptide associated with eggs causes a mobility shift in a major plasma membrane protein of spermatozoa. J. Biol. Chem. 259:14874–14879

Takayanagi, R., Inagami, T., Snajdar, R.M., Imada, T., Tamura, M. & Misono, K.S. (1987) Two distinct forms of receptors for atrial natriuretic factor in adrenocortical cells. J. Biol. Chem. 262:12104–12113

Thorpe, D.S. & Garbers, D.L. (1989) The membrane form of guanylate cyclase: homology with a subunit of the cytoplasmic form of the enzyme. J. Biol. Chem. 264:6545–6549

Vacquier, V.D. & Moy, G.W. (1986) Stoichiometry of phosphate loss from sea urchin sperm guanylate cyclase during fertilization. Biochem. Biophys. Res. Commun. 137:1148–1152

Vandlen, R.L., Arcuri, K.E. & Napier, M.A. (1985) Identification of a receptor for atrial natriuretic factor in rabbit aorta membranes by affinity cross-linking. J. Biol. Chem. 260:10889–10892

Waldman, S.A., Rapoport, R.M. & Murad, F. (1984) Atrial natriuretic factor selectively activates particulate guanylate cyclase and elevates cyclic GMP in rat tissues. J. Biol. Chem. 259:14332–14334
Ward, G.E. & Vacquier, V.D. (1983) Dephosphorylation of a major sperm membrane protein is induced by egg jelly during sea urchin fertilization. Proc. Natl. Acad. Sci. USA 80:5578–5582
Ward, G.E., Garbers, D.L. & Vacquier, V.D. (1985) Effects of extracellular egg factors on sperm guanylate cyclase. Science 227:768–770
White, A.A. & Aurbach, G.D. (1969) Detection of guanyl cyclase in mammalian tissues. Biochim. Biophys. Acta 191:686–697
Yip, C.C., Laing, L.P. & Flynn, T.G. (1985) Photoaffinity labeling of atrial natriuretic factor receptors of rat kidney cortex plasma membranes. J. Biol. Chem. 260:8229–8232

Mechanisms for Hormonal Regulation of the Different Isoforms of Guanylate Cyclase

F. MURAD[1]

1 Introduction

Cyclic GMP was first described in urine and other biological samples more than 25 years ago. The description and characterization of the enzymes responsible for cyclic GMP synthesis (guanylate cyclase), cyclic GMP hydrolysis (phosphodiesterase), and expression of some of the effects of cyclic GMP (cyclic GMP-dependent protein kinase) followed shortly thereafter. It soon became apparent that various hormones could increase cyclic GMP synthesis and accumulation in different tissues. Thus, it was anticipated that cyclic GMP would serve as a second messenger of many hormone-induced responses (see Murad 1986; Murad et al. 1978, 1979, 1988; Waldman and Murad 1987). However, the mechanisms of hormone-induced cyclic GMP accumulation in tissues and the coupling of hormone receptors to cyclic GMP synthesis have remained unresolved until the past several years. The major difficulties in answering these interesting questions are due to the multiple isoenzyme forms of guanylate cyclase, the diverse mechanisms of hormone-receptor coupling to the regulation of each of these isoenzymes, and the failure of hormones to activate the enzyme in cell-free preparations. The detailed characterization and purification of these isoforms of guanylate cyclase in our laboratory and others have permitted us to begin to understand the mechanisms of hormonal regulation of the guanylate cyclase isoenzyme family. Some of the properties of the isoforms of guanylate cyclase and the mechanisms of hormone-receptor coupling to each of these isoforms will be briefly reviewed here. Interestingly, the four isoenzyme forms of guanylate cyclase that have been described to date are each uniquely coupled to hormonal regulation. The evolutionary need for such diverse regulation of this enzyme family and cyclic GMP synthesis is not readily apparent. However, the presence of both a hormone receptor and guanylate cyclase catalytic activity in the same transmembrane macromolecule of one isoform may suggest that the cyclic GMP system represents a very primitive evolutionary mechanism of hormone signal transduction. Presumably, this system evolved further into several macromolecules (receptors, transducers, and cyclases) that interact through different complex signal transduction pathways. This diverse system of hormone-cyclase coupling is best represented in the effects of several classes of vasodilators which will also be discussed.

[1]Abbott Laboratories, Abbott Park, Illinois 60064, USA

40. Colloquium Mosbach 1989
Molecular Mechanisms of Hormone Action

2 Results and Discussion

To date, four isoenzyme forms of guanylate cyclase have been described and characterized (Table 1). Three of these isoforms have been purified: the soluble or cytosolic form (Kamisaki et al. 1986), the membrane-associated form possessing the ANF-R_1 receptor (Kuno et al. 1986), and the membrane-associated form with no ANF binding (Waldman et al. 1989). The isoform associated with the cytoskeletal or microtubular elements has been partially purified and characterized and its detailed structure and properties should be forthcoming in the near future (Waldman and Murad 1986; and our unpublished observations). For detailed reviews of these isoforms, readers are also referred to some of our recent review articles (Murad 1986, 1988; Murad et al. 1988; Waldman and Murad 1987). One of the membrane-associated isoforms of guanylate cyclase possessing one of the ANF receptors (ANF-R_1) has many similarities to the membrane form from sea urchin sperm which is activated by egg peptide chemoattractant factors. This isoform is described in greater detail in Chapter 19 (this Vol.).

Some of the hormones, toxins, peptides, and ligands that couple to and activate each of these isoenzyme forms are summarized in Table 1. Many hormones such as acetylcholine, histamine, substance P, oxytocin, etc. can increase cyclic GMP synthesis in intact cells and tissues but not in cell-free preparations. These hormones typically

Table 1. Isoforms of guanylate cyclases and their properties and methods of regulation (After Murad 1988; Murad et al. 1988)

	Isoform	Structure	Activators	Mechanism of hormone-receptor coupling
I.	*Soluble*			
a)	Cytosolic	150 kDa heterodimer of 82 + 70 kDa subunits	Acetylcholine, histamine, sub. P, oxytocin, etc.	Formation of EDRF/NO and perhaps other free radicals
II.	*Particulate*			
a)	Membrane (associated with ANF or egg factor receptors)	130 kDa transmembrane glycoprotein	Natriuretic peptides (ANF,BNF), egg factors, ? others	Protein factor requiring ATP
b)	Membrane (no apparent receptor associated as yet)	130 kDa transmembrane glycoprotein	Unknown at present, (probably a unique receptor ligand)	Unknown (probably an ATP-requiring protein
c)	Cytoskeletal	Unknown (probably a 60–70 kDa multimeric protein)	*E. coli* heat-stable enterotoxin (ST), ? similar endogenous peptides	Protein factor requiring ATP or GTP

require extracellular calcium and their effects are blocked with agents such as methylene blue which prevents the activation of the soluble isoform of guanylate cyclase (Katsuki et al. 1977; Leitman 1988; Waldman and Murad 1987; Ishii et al. 1989). These hormones and agents fail to activate guanylate cyclase in cell-free preparations, presumably because the complex mechanisms required for hormone-receptor coupling to soluble guanylate cyclase activation have been disrupted. We have recently shown that those receptors which couple to soluble guanylate cyclase activation increase the formation of an endogenous nitrovasodilator which liberates nitric oxide free radical (Ishii et al. 1989; unpublished observations). This signal transduction mechanism became apparent from our earlier work with endothelial-derived relaxant factor (EDRF) effects on cyclic GMP accumulation in vascular preparations (Rapoport and Murad 1983). We have since found that EDRF can also be generated by LLC-PK_1 cells (Ishii et al. 1989), as well as endothelial cells. In LLC-PK_1 cells, we have also found that hormone effects on cyclic GMP accumulation were similar to the effects of sodium nitroprusside and nitric oxide in that they were prevented with methylene blue (Leitman et al. 1988). From our work and the work of others (see below), it is now apparent that numerous cell types can produce EDRF in addition to endothelial cells (see Fig. 1). The formation and/or release of EDRF that is provoked by numerous hormones and ligands is calcium- and energy-dependent. The recent work of Moncada et al. (1988) suggests that arginine or a related compound can be converted to EDRF which they and others (Ignarro 1987; Furchgott 1988) have proposed to be the nitric oxide free radical. It is probably very unlikely that EDRF is itself the nitric oxide free radical since such a reactive material would be very labile as a paracrine substance or autocoid in the interstitial space or as it traverses cell

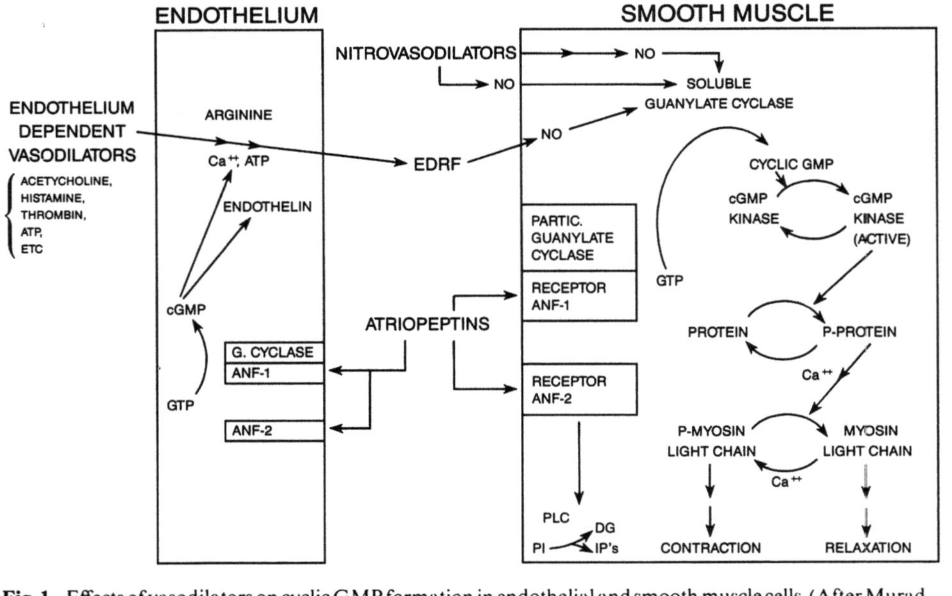

Fig. 1. Effects of vasodilators on cyclic GMP formation in endothelial and smooth muscle cells. (After Murad 1986)

membranes. A more appealing hypothesis is that some classes of hormones (acetylcholine, histamine, substance P, oxytocin, etc.) increase the formation of an "EDRF-like" material from an arginine-like precursor and this reactive substance (EDRF) can be viewed as an "endogenous nitrovasodilator" as we suggested previously (Murad 1986). This EDRF-like substance can probably be produced in most, if not all, cell types and can be both liberated from cells to act as a paracrine material to increase cyclic GMP synthesis in neighboring cells and can also act within the cell in which it is generated to activate soluble guanylate cyclase and increase cyclic GMP synthesis (see Fig. 2). Since proposing this hypothesis last year (Murad 1988; Murad et al. 1988), we have obtained a considerable amount of data to support this concept (Ishii et al. 1989; our unpublished observations).

The first report that arginine and related materials could activate soluble guanylate cyclase came from the work of Deguchi and Yoshioko (1982). They found that arginine activation of soluble guanylate cyclase required some unknown factor in their partially purified extracts from neuroblastoma cells. We confirmed their observations and found that arginine failed to activate purified or partially purified soluble guanylate cyclase, presumably because the required factor(s) were removed or destroyed during purification (our unpublished observations). Hibbs et al. (1987) have since reported that macrophages can generate nitric oxide from arginine. We have known for some time (Katsuki et al. 1977; Arnold et al. 1977) that nitric oxide formation from nitrovasodilators was the mechanism of increased cyclic GMP formation with these agents in various cells and tissues (see Murad and Aurbach 1978; Rapoport and Murad 1983; Murad 1986). We have also proposed that hormonally

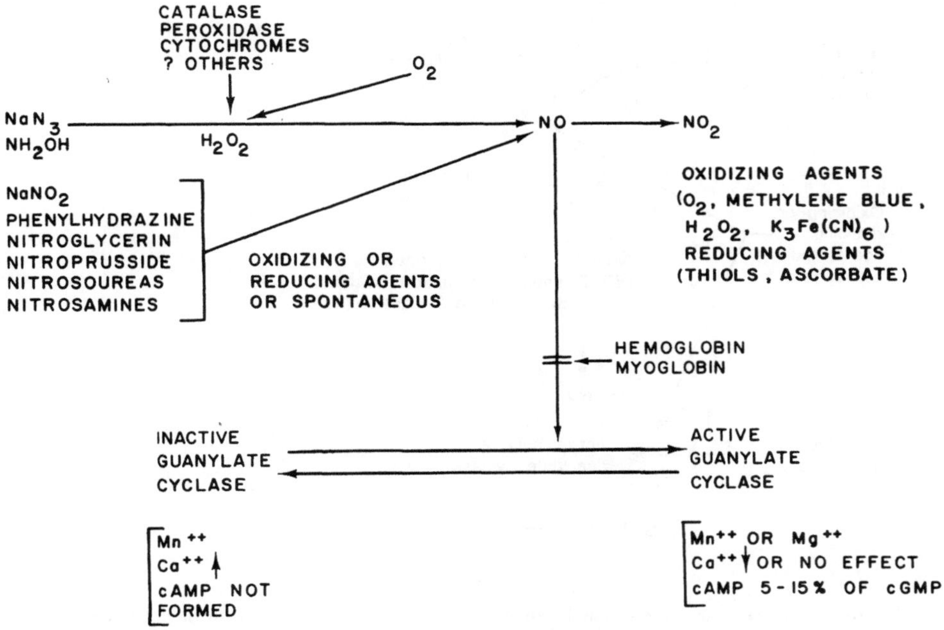

Fig. 2. Effects of nitrovasodilators and NO on guanylate cyclase activation (Murad et al. 1978)

induced cyclic GMP accumulation in tissues could be due to hormonal effects on the formation of reactive free radicals such as nitric oxide or other reactive species of oxygen (Murad et al. 1978, 1979; Mittal and Murad 1977).

The recent studies of Hibbs et al. (1987) and Moncada et al. (1988) prompted us to reexamine this hypothesis. As noted above, we now have substantial data to support the hypothesis that some classes of hormones alter the metabolism of arginine or a related substance to a reactive intermediate that is similar, if not identical, to EDRF (Ishii et al. 1989; our unpublished observations). The formation of this endogenous nitrovasodilator then results in the formation of nitric oxide, the activation of soluble guanylate cyclase, and increased cyclic GMP accumulation. These latter effects can occur in the same cell in which EDRF is formed (Ishii et al. 1989) or in adjacent neighboring cells (Rapoport et al. 1983). Many agents can interrupt the formation of EDRF and nitric oxide such as oxidizing and reducing agents [methylene blue, thiols, $K_3Fe(CN)_6$], some arginine analogs, which are competitive or noncompetitive inhibitors of this pathway, and sinks or scavengers of nitric oxide (hemoglobin, myoglobin, etc.) (Fig. 2; see reviews for earlier proposed models to interrupt NO formation and guanylate cyclase activation, Murad et al. 1978, 1979). Nonpermeable compounds such as hemoglobin or myoglobin prevent EDRF/NO effects in neighboring adjacent cells, while permeable compounds can prevent hormone effects in the cells in which EDRF/NO is generated. This scheme is summarized in Figs. 2 and 3.

The mechanisms and agents which alter the formation of nitric oxide from various nitrovasodilators or scavenge the NO and prevent soluble guanylate cyclase activation (Fig. 2) are remarkably similar to the agents that disrupt the hormonal coupling to increased cyclic GMP synthesis via the soluble isoform of guanylate cyclase (Fig. 3).

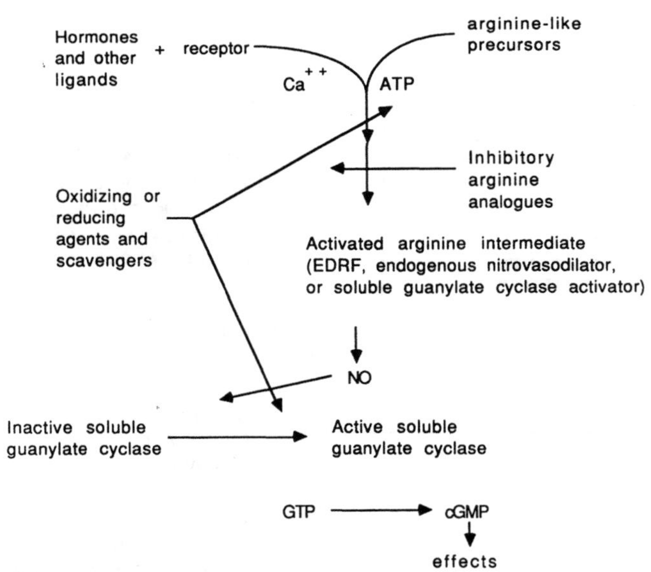

Fig. 3. Proposed mechanisms of hormonal activation of soluble guanylate cyclase through EDRF and NO formation. EDRF and/or NO may also be released from the cell in which it is generated to increase cyclic GMP in neighboring and adjacent cells

The pathways proposed in Fig. 3 explain many of the observations we and others have made with regard to hormonal regulation of cyclic GMP synthesis. Hormones that are coupled to the activation of soluble guanylate cyclase and cyclic GMP synthesis require intact cells and external calcium ion (see Katsuki et al. 1977; Bohme et al. 1978), presumably because of the complex signal transduction cascade that requires the formation of the EDRF-like substance from an arginine-like precursor. The formation of EDRF is presumably Ca^{2+}- and ATP-dependent. The characterization of this enzyme pathway and the substrates and cofactors required should become apparent in the near future. Modifications of this pathway are expected to provide important new mechanisms of amplifying or decreasing signal transduction via this pathway.

Methylene blue, an inhibitor of the activation of soluble guanylate cyclase (Katsuki et al. 1977) can also prevent the effects of some hormones on cyclic GMP accumulation that are mediated through the activation of soluble guanylate cyclase (Leitman et al. 1988). Hormones such as ANF and toxins such as *E. Coli* enterotoxin that are coupled to the activation of particulate isoforms of guanylate cyclase do not require Ca^{2+} (Hughes 1978; Field 1978; Guerrant et al. 1980; Winquist et al. 1984; and Waldman et al. 1984) and are not inhibited by methylene blue (Leitman et al. 1988; unpublished observations). Some macromolecules which are not permeable to cell membranes such as hemoglobin or myoglobin can prevent the activation of soluble guanylate cyclase in cell-free preparations by nitric oxide and various nitrovasodilators (Mittal et al. 1978) and can prevent the effects of endothelium-dependent vasodilators due to the inactivation of EDRF and/or nitric oxide in the extracellular space of blood vessels (Furchgott 1988). Thus, hormonal regulation of soluble guanylate cyclase can occur within the same cell through the generation of nitric oxide or perhaps other free radicals as we postulated many years ago (Murad et al. 1978, 1979). This signal transduction mechanism can also be utilized between neighboring cells as is the case with endothelial-dependent vasodilators (Rapoport and Murad 1983). In both paradigms, nitric oxide is participating as an intracellular second messenger and the precursor of NO which is probably EDRF can act within the same cell in which it is generated as well as a paracrine substance. Agents that prevent the formation of EDRF and NO or the activation of soluble guanylate cyclase should predictably block hormone effects that are mediated through activation of soluble guanylate cyclase and cyclic GMP. We are continuing to examine the effects of numerous classes of such inhibitors. We would also propose that EDRF is too restrictive a name for a material that activates soluble guanylate cyclase and is generated in numerous cell types. We would suggest that EDRF be called the "soluble guanylate cyclase activator" which couples some classes of hormone receptors to soluble guanylate cyclase activation in a complex signal-transduction process.

Other classes of hormones such as natriuretic factors (atrial natriuretic peptide, brain natriuretic peptide, etc.) are coupled to the activation of the membrane-associated guanylate cyclase (Winquist et al. 1984; Waldman et al. 1984; Song et al. 1988). Indeed, the ANF-R_1 receptor which is coupled to cyclic GMP synthesis copurifies with the membrane isoenzyme form of guanylate cyclase (Kuno et al. 1986). We have proposed that this transmembrane glycoprotein possesses an ANF-R_1 receptor in the extracellular domain and guanylate cyclase catalytic activity in the intracellular domain (Kuno et al. 1986). The cDNA for this protein has been recently cloned from a rat brain library and expressed (Chinkers et al. 1989). When expressed in cos cells, both ANF binding and guanylate cyclase activity were obtained. Although both activities reside in the same transmembrane macromolecule, ANF activation of guanylate cyclase is functionally uncoupled in purified preparations (Kuno et al. 1988). In crude preparations ANF activation of guanylate cyclase is markedly potentiated with the addition of adenine nucleotides (Song et al. 1988; Chang et al. 1989). We have obtained evidence that a protein factor is required in this signal-transduction cascade and that a protein phosphorylation event is participating in the coupling process (Chang et al. 1989; our unpublished observations).

We have also purified a membrane-associated isoform of guanylate cyclase which does not possess ANF-R_1 receptor sites (Waldman et al. 1989). This isoform has very similar properties to the isoform with ANF-R_1 receptors and may possess receptor sites for hormones or classes of hormones that are yet to be described.

E. Coli heat-stable enterotoxin (ST) is also coupled to a particulate isoform of guanylate cyclase (Guerrant et al. 1980; Waldman et al. 1986; Kuno et al. 1986). The isoform of guanylate cyclase that is coupled to ST is probably the cytoskeletal form (Waldman et al. 1986, 1989). The ST receptor and cytoskeletal or membrane guanylate cyclases are different macromolecules that are readily separated with different techniques (Kuno et al. 1986). The ST receptor and cytoskeletal guanylate cyclase also appear to be coupled by processes that require nucleoside triphosphates and protein factors (our unpublished observations).

3 Summary

To date, four (or perhaps more) isoforms of guanylate cyclase are present in mammalian tissues. Each of these isoforms is uniquely coupled to hormonal regulation of cyclic GMP synthesis. With the soluble isoform of guanylate cyclase, some classes of hormones increase the formation of an activator that resembles, or is identical to, EDRF which gives rise to nitric oxide. Thus, with this signal-transduction system, the nitric oxide free radical is the second messenger and cyclic GMP is the third messenger. This EDRF-like material can act within the cell in which it is generated or can act as a paracrine material to increase cyclic GMP in neighboring cells via the formation of NO and the activation of soluble guanylate cyclase. There are at least three isoforms of particulate guanylate cyclase, two transmembrane glycoproteins and a cytoskeletal form. One of the membrane isoforms is also the ANF-R_1 receptor. Presumably other membrane-associated isoforms will be receptors for other hormones which are yet to be identified. Receptors for *E. Coli* heat-stable enterotoxin (ST) are probably coupled to the cytoskeletal isoform of guanylate cyclase. The particulate isoforms of guanylate cyclase are also coupled to hormonal regulation via complex reactions that require the participation of nucleoside triphosphates, protein factors, and protein phosphorylation.

Acknowledgments. The studies and experiments that have led to the hypotheses in this review have been conducted over many years with the collaboration of many outstanding associates and trainees and with research support from the NIH, Abbott Laboratories, and many other agencies.

References

Arnold, W.P., Mittal, C.K., Katuski, S. & Murad, F. (1977) Nitric oxide activates guanylate cyclase and increases guanosine 3',5'-monophosphate levels in various tissue preparations. Proc. Natl. Acad. Sci. USA 74:3203-3207

Bohme, E., Graf, H. & Schultz, G. (1978) Effects of sodium nitroprusside and other smooth muscle relaxants on cyclic GMP formation in smooth muscle and platelets. Adv. Cyclic Nucl. Res. 9:131-143

Chang, C.H., Kohse, K., Chang, B. & Murad, F. (1989) Participation of protein phosphorylation in the activation of particulate guanylate cyclase by ANP. FASEB Vol 3, Issue 4, A1005

Chinkers, M., Garbers, D., Chang, M.S., Lowe, D., Chin, H., Goeddel, D.V. & Schulz, S. (1989) A membrane form of guanylate cyclase is an atrial natriuretic peptide receptor. Nature (London) 338:78-83

Deguchi, T. & Yoshioko, M. (1982) L-Arginine identified as an endogenous activator for soluble guanylate cyclase from neuroblastoma cells. J. Biol. Chem. 257:10147-10151

Field, M., Graf, L.H., Laird, W.J. & Smith, P.L. (1978) Heat-stable enterotoxin of *E. coli*: in-vitro effects on guanylate cyclase activity, cyclic GMP concentration and ion transport in small intestine. Proc. Natl. Acad. Sci. USA 75:2800-2804

Furchgott, R.F. (1988) Studies on the relaxation of rabbit aorta by sodium nitrite: the basis for the proposal that the acid-activatable inhibitory factor from bovine retractor penis is inorganic nitrite and EDRF is nitric oxide. In: VanHoutte, P. (ed.) Mechanisms of vasodilation. Raven Press, New York 401-414

Guerrant, R.L., Hughes, J.M., Chang, B., Robertson, D.C. & Murad, F. (1980) Activation of intestinal guanylate cyclase by heat stable enterotoxin of *Escherichia coli*: studies of tissue specificity, potential receptors and intermediates. J. Infect. Dis. 142:220-228

Hibbs, J.D., Vavrin, Z. & Taintor, R.L. (1987) L-arginine is required for expression of the activated macrophage effector mechanism causing selective metabolic inhibition in target cells. J. Immunol. 138:550-565

Hughes, J., Murad, F., Chang, B. & Guerrant, R. (1978) The role of cyclic GMP in the mechanism of action of heat-stable enterotoxin of *E. coli*. Nature (London) 271:755-756

Ignarro, L.J., Buga, G.M., Wood, K.S., Byrns, R.E. & Chaudhuri, G. (1987) Endothelium-derived relaxing factor produced and released from artery and vein is nitric oxide. Proc. Natl. Acad. Sci. USA 84:9265-9269

Ishii, K., Gorsky, L., Förstermann, V. & Murad, F. (1989) Endothelium-derived relaxing factor: the endogenous activator of soluble guanylate cyclase in various types of cells. J. Applied Cardiology (in press)

Kamisaki, Y., Saheki, S., Nakane, M., Palmieri, J., Kuno, T., Chang, B., Waldman, S.A. & Murad, F. (1986) Soluble guanylate cyclase from rat lung exists as a heterodimer. J. Biol. Chem. 261:7236-7241

Katsuki, S. & Murad, F. (1977) Regulation of cyclic 3',5'-adenosine monophosphate and cyclic 3',5'-guanosine monophosphate levels and contractility in bovine tracheal smooth muscle. Mol. Pharmacol. 13:330-341

Katsuki, S., Arnold, W., Mittal, C.K. & Murad, F. (1977) Stimulation of guanylate cyclase by sodium nitroprusside, nitroglycerin and nitric oxide in various tissue preparations and comparison to the effects of sodium azide and hydroxylamine. J. Cyclic Nucl. Res. 3:23-25

Kuno, T., Andresen, J.W., Kamisaki, Y., Waldman, S.A., Chang, L.Y., Saheki, S., Leitman, D.C., Nakane, M. & Murad, F. (1986) Co-purification of an atrial natriuretic factor receptor and particulate guanylate cyclase from rat lung. J. Biol. Chem. 261:5817-5823

Kurose, H., Inagami, T. & Ui, M. (1987) Participation of adenosine 5' triphosphate in the activation of membrane-bound guanylate cyclase by the atrial natriuretic factor. FEBS Lett. 219:375-379

Leitman, D.C., Agnost, V.L., Catalano, R.M., Schroeder, H., Waldman, S.A., Bennett, B.M., Tuan, J.J. & Murad, F. (1988) Atrial natriuretic peptide, oxytocin and vasopressin increase cyclic GMP in LLC-PK_1 kidney epithelial cells. Endocrinology 122:1478-1485

Mittal, C.K. & Murad, F. (1977) Properties and oxidative regulation of guanylate cyclase. J. Cyclic Nucl. Res. 3:381-391

Mittal, C.K., Arnold, W.P. & Murad, F. (1978) Characterization of protein inhibitors of guanylate cyclase activation from rat heart and bovine lung. J. Biol. Chem. 253:1266-1271

Moncada, S., Palmer, R.M.J. & Higgs, E.A. (1988) The discovery of nitric oxide as the endogenous nitrovasodilator. Hypertension 12:365-372

Murad, F. (1986) Cyclic guanosine monophosphate as a mediator of vasodilation. J. Clin. Invest. 78:1-5

Murad, F. (1989) Modulation of the guanylate cyclase-cyclic GMP system by vasodilators and the role of free radicals as second messengers. Proc. NATO Adv. Stud. Inst. Vascular endothelium: receptors and transduction mechanisms, Porto Carros, Greece June, 1988, in *Vascular Endothelium* eds J.D. Catrovos, C.N. Gillis and U.S. Ryan, Plenum Pub. 157–164

Murad, F. & Aurbach, G.D. (1978) Cyclic GMP in metabolism: interrelationship of biogenic amines, hormones and other agents. In: Freinkel, N. (ed.) The year in metabolism 1976–1977. Plenum, New York, pp 1–32

Murad, F., Mittal, C.K., Arnold, W.P., Katsuki, S. & Kimura, H. (1978) Guanylate cyclase: activation by azide, nitro compounds, nitric oxide, and hydroxyl radical and inhibition by hemoglobin and myoglobin. Adv. Cyclic Nucl. Res. 9:145–158

Murad, F., Arnold, W.P., Mittal, C.K. & Braughler, J.M. (1979) Properties and regulation of guanylate cyclase and some proposed functions for cyclic GMP. Adv. Cyclic Nucl. Res. 11:175–204

Murad, F., Leitman, D., Waldman, S., Chang, C.H., Hirata, M. & Kohse, K. (1988) Effects of nitrovasodilators, endothelium-dependent vasodilators and atrial peptides on cGMP. Cold Spring Harbor Symp. Quant. Biol. 53:1005–1009

Rapoport, R.M. & Murad, F. (1983) Endothelium-dependent and nitrovasodilator-induced relaxation of vascular smooth muscle: role for cyclic GMP. J. Cyclic Nucl. Protein Phosphor. Res. 9:281–296

Song, D.L., Kohse, K. & Murad, F. (1988) Brain natriuretic factor: augmentation of cellular cyclic GMP, activation of particulate guanylate cyclase and receptor binding. FEBS Lett. 1:125–129

Waldman, S.A. & Murad, F. (1987) Cyclic GMP synthesis and function. Pharm. Rev. 39:163–196

Waldman, S.A., Rapoport, R.M. & Murad, F. (1984) Atrial natriuretic factor selectively activates particulate guanylate cyclase and elevates cyclic GMP in rat tissues. J. Biol. Chem. 259:14332–14334

Waldman, S.A., Kuno, T., Kamisaki, Y., Chang, L.Y., Garieppy, J., O'Hanley, P.D., Schoolnik, G.K. & Murad, F. (1986) Intestinal receptor for heat-stable enterotoxin of *E. Coli* is tightly coupled to a novel form of particulate guanylate cyclase. Inf. Immunol. 51:320–326

Waldman, S.A., Leitman, D.C., Chang, L.Y. & Murad, F. (1989) Comparison of particulate guanylate cyclase in cells with and without atrial natriuretic peptide receptor binding activity Mol. Cell. Biochem. (in press).

Winquist, R.M., Faison, E.P., Waldman, S.A., Schwartz, K., Murad, F. & Rapoport, R.M. (1984) Atrial natriuretic factor elicits an endothelium independent relaxation and activates particulate guanylate cyclase in vascular smooth muscle. Proc. Natl. Acad. Sci. 81:7661–7664

Catecholamine Receptors: Structure, Function, and Regulation

M.G. CARON, B.K. KOBILKA, M. BOUVIER, W.P. HAUSDORFF, J.L. BENOVIC, M. LOHSE, S. COTECCHIA, J.W. REGAN, A. FARGIN, J.R. RAYMOND, and R.J. LEFKOWITZ[1]

1 Introduction

The various receptors for catecholamines, termed adrenergic receptors, represent excellent model systems for the study of receptor-mediated transmembrane signaling systems because of their ubiquity, coupling to well-defined effector mechanisms, and the clinical importance of drugs which interact with them. The β_1 and β_2-adrenergic receptors stimulate adenylyl cyclase via the guanine nucleotide regulatory protein G_s. The α_2-adrenergic receptors inhibit adenylyl cyclase via G_i. The α_1-adrenergic receptors stimulate hydrolysis of polyphosphoinositides by activating phospholipase C, thus generating inositol triphosphate and diacylglycerol. Each of these systems is in turn analogous to the retinal light transduction system which consists of the prototypic receptor rhodopsin, a G-protein transducin, and an effector enzyme, which is a cyclic GMP phosphodiesterase.

2 Results and Discussion

2.1 Structure and Properties of Adrenergic Receptors

Over the past few years our group has succeeded in cloning the genes and/or cDNAs for all of the known adrenergic receptors (Lefkowitz and Caron 1988). We have, in addition, cloned two other related genes/cDNAs; one for a previously unsuspected α_2-adrenergic receptor (Regan et al. 1988), and the other for the genomic clone G-21 which encodes the serotonin 5HT1A receptor (Fargin et al. 1988). Our ability to clone these genes was based on our having developed means for purifying to homogeneity the various subtypes of adrenergic receptors. With pure proteins we were able to obtain limited protein sequence information, design oligonucleotide probes, and clone the genes and/or cDNAs for these proteins. An alignment of the sequences for the human β_1, β_2, α_2-C10, α_2-C4, and hamster α_1-adrenergic receptors is presented in Fig. 1.

Some of the information that has emerged from the deduced sequences of the various adrenergic receptors is summarized in Table 1. The purified proteins all consist of single polypeptide chains of molecular weights 64 000–80 000 daltons. They range in size from 413 amino acids for the human β_2-adrenergic receptor to 515 amino acids for the hamster α_1-adrenergic receptor. All of these receptors share the characteristic

[1]Depts. of Cell Biology, Biochemistry and Medicine, Howard Hughes Medical Institute, Duke University Medical Center, Durham, NC 27710, USA

40. Colloquium Mosbach 1989
Molecular Mechanisms of Hormone Action

```
                                                                          M-I
HUM β2AR:     --------------------------MGQPGNGSAFLLAPNAPDHDVTQRSHQRDEVWVVGMGIVMSLIVLAIVFGNVLV  54
HUM β1AR:     -MGAGVLVLGASEPGNLSSAAPLPDGAATAARLLVPASPPASLLPPASESPEPLSQQWTAGMGLLMALIVLLIVAGNVLV  79
HAM α1AR:     --------------MNPDLDTGHNTSAPAQWGELKDANFTGPNQTSSNSTLPQLDVTRAISVGLVLGAFILFAIVGNILV  66
HUM α2AR-C10: --------------------------MGSLQPDAGNASWNGTEAPGGGARATPYSLQVTLTLVCLAGLLMLLTVFGNVLV  54
HUM α2AR-C4:  --------MASPALAAALAVAAAAGPNASGAGERGSGGVANASGASWGPPRGQYSAGAVAGLAAVVGFLIVFTVVGNVLV  72

                          M-II                               M-III
HUM β2AR:     ITAIAKFERLQTVTNYFITSLACADLVMGLAVVPFGAAHILMKMWTFGNFWCEFWTSIDVLCVTASIETLCVIAVDRYFA  134
HUM β1AR:     IVAIAKTPRLQTLTNLFIMSLASADLVMGLLVVPFGATIVVWGRWEYGSFFCELWTSVDVLCVTASIETLCVIALDRYLA  159
HAM α1AR:     ILSVACNRHLRTPTNYFIVNLAIADLLLSFTVLPFSATLEVLGYWVLGRIFCDIWAAVDVLCCTASILSLCAISIDRYIG  146
HUM α2AR-C10: IIAVFTSRALKAPQNLFLVSLASADILVATLVIPFSLANEVMGYWYFGKTWCEIYLALDVLFCTSSIVHLCAISLDRYWS  134
HUM α2AR-C4:  VIAVLTSRALRAPQNLFLVSLAAADILVATLVMPFSLANELMAYWYFGQVWCGVYLALDVLFCTSSIVHLCAISLDRYWS  152

                         M-IV                                                 M-V
HUM β2AR:     ITSPFKYQSLLTKNKARVIILMVWIVSGLTSFLPIQMHWYRATHQEAINCYANETCCDFFTNQAYAIASSIVSFYVPLVI  214
HUM β1AR:     ITSPFRYQSLLTRARARGLVCTVWAISALVSFLPILMHWWRAESDEARRCYNDPKCCDFVTNRAYAIASSVVSFYVPLCI  239
HAM α1AR:     VRYSLQYPTLVTRRKAILALLSVWVLSTVISIGPLLGWKEPAPNDD--------KECGVTEEPFYALFSSLGSFYIPLAV  218
HUM α2AR-C10: ITQAIEYNLKRTPRRIKAIIITVWVISAVISFPPLISIEKKGGGGGP---QPAEPRCEINDQKWYVISSCIGSFFAPCLI  211
HUM α2AR-C4:  VTQAVEYNLKRTPRRVKATIVAVWLISAVISFPPLVSLYRQPDGAAY-------PQCGLNDETWYILSSCIGSFFAPCLI  225

HUM β2AR:     MVFVYSRVFQEAKRQLQKIDKSEGRFHVQNLSQ-----------------------------------------------  247
HUM β1AR:     MAFVYLRVFREAQKQVKKIDSCERRFLGGPARPPSPSPSPVPAPAP----------------------------------  285
HAM α1AR:     ILVMYCRVYIVAKRTTKNLEAGVMKEMSNSKELTLRIHSKN---------------------------------------  259
HUM α2AR-C10: MILVYVRIYQIAKRRTRVPPSRRGPDAVAAPPGGTERRPNGLGPERSAGPGGAEAEPLPTQLNGAPGEPAPAGPRDTDAL  291
HUM α2AR-C4:  MGLVYARIYRVAKRRTRTLSEKRAPVGPDGASPTTENGLGAAAGEARTGTARPRPPTWSRTRAAQRPRGGAPGPLRRGGR  305

HUM β2AR:     ---------------------------------------------------------VEQDGRTGHGLRRSSKFCLKEHK  270
HUM β1AR:     --------------------------------------------PPGPPRPAAAAATAPLANGRAGKRRPSRLVALREQK  321
HAM α1AR:     ------------------------------------------------FHEDTLSSTKAKGHNPRSSIAVKLFKFSREKK  291
HUM α2AR-C10: DLEESSSSDHAERPPGPRRPERGPRGKGKARASQVKPGDSLRGAGRGRRGSGRRLQGRGRSASGLPRRRAGAGGQNREKR  371
HUM α2AR-C4:  -------RRAGAEGGAGGADGQGAGPGAAQSGALTASRSPGPGGRLSRASSRSVEFFLSRRRRARSSVCRRKVAQAREKR  378

                       M-VI                          M-VII
HUM β2AR:     ALKTLGIIMGTFTLCWLPFFIVNIVHVIQDNL--IRKEVYILLNWIGYVNSGFNPLIYCRS-PDFRIAFQELL-CLRRSS  346
HUM β1AR:     ALKTLGIIMGVFTLCWLPFFLANVVKAFHREL--VPDRLFVFFNWLGYANSAFNPIIYCRS-PDFRKAFQGLLCCARRAA  398
HAM α1AR:     AAKTLGIVVGMFILCWLPFFIALPLGSLFSTLKP-PDAVFKVVFWLGYFNSCLNPIIYPCSSKEFKRAFMRILGCQCRSG  370
HUM α2AR-C10: FTFVLAVVIGVFVVCWFPFFFTYTLTAVGCS---VPRTLFKFFFWFGYCNSSLNPVIYTIFNHDFRRAFKKIL-CRGDRK  447
HUM α2AR-C4:  FTFVLAVVMGVFVLCWFPFFFIYSLYGICREACQVPGPLFKFFFWIGYCNSSLNPVIYTVFNQDFRPSFKHIL-FRRRRR  457

HUM β2AR:     LKAYGNGYSSNGNTGEQSGYHVEQEKENKLLCEDLPGTEDFVGHQGTVPSDNIDSQGRNCSTNDSLL:------------  413
HUM β1AR:     RRRHATHGDRPRASGCLARPGPPPSPGAASDDDDDDVVGATPPARLLEPWAGCNGGAAADSDSSLDEPCRPGFASESKV:  477
HAM α1AR:     RRRRRRRRLGACAYTYRPWTRGGSLERSQSRKDSLDDSGSCMSGSQRTLPSASPSPGYLGRGAQPPLELCAYPEWKSGAL  450
HUM α2AR-C10: RIV:----------------------------------------------------------------------------  450
HUM α2AR-C4:  GFRQ:---------------------------------------------------------------------------  461

HUM β2AR:     ------------------------------------------------------------------  413
HUM β1AR:     ------------------------------------------------------------------  477
HAM α1AR:     LSLPEPPGRRGRLDSGPLFTFKLLGEPESPGTEGDASNGGCDATTDLANGQPGFKSNMPLAPGHF:  515
HUM α2AR-C10: ------------------------------------------------------------------  450
HUM α2AR-C4:  ------------------------------------------------------------------  461
```

Table 1. Properties of the cloned adrenergic receptors

Subtype (species)	M_r-SDS-PAGE	Peptide length	Introns	% Identity[a]
		aa		
β_2 (Human)	64 000	413	0	100
β_1 (Human)	64 000	477	0	71
α_{2A}-C10 (Human)	64 000	450	0	42
α_{2B}-C4 (Human)	75 000	461	?	39
α_1 (Hamster)	80 000	515	1	42

[a] Sequence identity within the presumed membrane spanning domains between a given receptor and the human β_2-adrenergic receptor. The α_2-C10 and α_2-C4 adrenergic receptors are, respectively, those purified from human platelet and cloned and cloned from a human kidney cDNA library. These receptors have also been referred to as α_{2A} and α_{2B}, respectively (Kobilka et al. 1987; Regan et al. 1988).

and now well-known feature of having seven clusters of hydrophobic amino acids, 20–28 residues in length, long enough to represent membrane spanning domains. The same topology is shared by rhodopsin.

The adrenergic receptors, as well as all members of the G-protein-coupled receptor family, share significant amino acid sequence similarities but more particularly within the presumed membrane spanning regions. Within these domains the adrenergic receptors generally share 40–50% sequence identity. Various studies have suggested that the ligand binding sites for these receptors reside within the pocket presumably formed by the various transmembrane segments coming together in the plasma membrane (Dohlman et al. 1988; Rubinstein et al. 1987; Dixon et al. 1987; Chung et al. 1988; Strader et al. 1987b, 1988; Kobilka et al. 1988). The only features of the receptors which vary significantly in size and composition are the amino terminus, the third cytoplasmic loop, and the carboxyl terminus. The latter two regions appear to be involved in receptor G-protein interactions. In our studies with chimeric receptors (Kobilka et al. 1988) and site-directed mutagenesis (O'Dowd et al. 1988) we have identified the amino and carboxyl terminal regions of the third cytoplasmic loop (adjacent to transmembrane segments 5 and 6) and the amino side of the C-terminal tail of the β_2-adrenergic receptor as being important for the interaction of the receptor with G_s. Similar results have been obtained by Strader, Dixon, and collaborators on the hamster β_2-adrenergic receptor (Dixon et al. 1987; Strader et al. 1987a).

Three of the five receptor genes listed in Table 1 have been found to lack introns within their coding regions. Only the α_1-adrenergic receptor contains a single intron within its coding block (Cotecchia et al. 1988). Rhodopsin contains several introns.

Fig. 1. Alignment of the sequences of the various cloned adrenergic receptors. The sequences shown are for the human β_1, β_2, α_2-C10, α_2-C-4 adrenergic receptors as well as the hamster α_1-adrenergic receptor. To the *right* of each line, the *number* corresponds to the amino acid number for each sequence. Below the dark lines and denoted *M-I* through *M-VII* are the presumed location of the seven membrane segments

2.1 Regulation of Receptor Function

Over the last several years the availability of specific tools with which receptors can be characterized has allowed the examination of mechanisms by which responsiveness to hormones and neurotransmitters can be regulated. Several mechanisms have been implicated in the dynamic regulation of responsiveness of these systems (Sibley et al. 1987; Clark 1986).

The paradigm we have examined in detail has been the agonist-mediated desensitization of the β_2-adrenergic stimulation of adenylyl cyclase. Desensitization is a general phenomenon of any hormone-responsive system and probably involves several distinct biochemical mechanisms. With respect to the receptor alone, uncoupling of the receptor from activation of the G-protein, sequestration of the receptor away from the plasma membrane as well as down-regulation of the receptor are mechanisms which can contribute to the development of this phenomenon. Covalent modification of the β_2-adrenergic receptor by various kinases has been suggested to play an important role in the uncoupling mechanism (Sibley et al. 1987). A novel kinase, the β-adrenergic receptor kinase (βARK), has been isolated and characterized which can phosphorylate the β_2-adrenergic receptor in an agonist-dependent fashion (Benovic et al. 1987).

Several lines of evidence suggest that phosphorylation of the receptor by this kinase plays an important role in the development of rapid agonist-mediated desensitization. First, β-adrenergic receptor kinase phosphorylation of pure β_2-adrenergic receptor in vitro reduces its ability to stimulate G_s. This impairment in function may involve a protein like arrestin which functions in the light transduction system to mediate the impairment of the ability of phosphorylated rhodopsin to activate transducin (Benovic et al. 1988). Second, mutation of the proposed carboxyl terminal βARK phosphorylation sites of the β_2-adrenergic receptor produces a dramatic change in the onset of rapid agonist-mediated desensitization. This change correlates with a loss in the ability of agonists to promote receptor phosphorylation in cells expressing these mutant receptors (Bouvier et al. 1988; Hausdorff et al. 1989). Third, in intact cells expressing these βARK phosphorylation site mutants, the rate of attenuation of the intracellular cAMP accumulation in response to an agonist is much slower than in cells expressing wild-type receptors (Liggett et al. 1989). Finally, in a permeabilized cell system, heparin which selectively inhibits βARK has been shown to inhibit both agonist-mediated rapid desensitization and phosphorylation of the β_2-adrenergic receptor with the same dose-response relationship (Lohse et al. 1989). These data, taken together, strongly suggest that the sites of phosphorylation of β_2-adrenergic receptor by the βARK residue on the carboxyl terminal of the protein and that the covalent modification of the receptor by βARK plays an important role in agonist-mediated desensitization.

3 Summary

The elucidation of the structure of the various adrenergic receptors indicates that they all share the seven transmembrane motif that is characteristic of G-protein-coupled receptors. The membrane spanning domains appear to be involved in determining agonist and antagonist specificity, whereas cytoplasmic regions of these receptors, in particular regions of the third cytoplasmic loop, may be responsible for interacting with G-proteins. Receptor-mediated signal transduction is a dynamically regulated process. The evidence reviewed here indicates that in the paradigm of agonist-mediated desensitization, the phosphorylation of the receptor by the β-adrenergic receptor kinase plays an important role. However, desensitization is a complex process involving other mechanisms such as receptor sequestration and at longer times down-regulation. In addition, the β_2-adrenergic receptor is phosphorylated by cAMP-dependent protein kinase in response to agonist occupancy (Benovic et al. 1985; Kunkel et al. 1989). An interesting prospect, with respect to agonist-mediated desensitization, will be the attempt to determine under which circumstances and physiological conditions these various mechanisms play a significant role.

References

Benovic JL, Pike LJ, Cerione RA, Staniszewski C, Yoshimasa T, Codina J, Birnbaumer L, Caron MG & Lefkowitz RJ (1985) Phosphorylation of the mammalian β-adrenergic receptor by cAMP-dependent protein kinase: regulation of the rate of receptor phosphorylation and dephosphorylation by agonist occupancy and effects on the coupling of the receptor to the stimulatory guanine nucleotide regulatory protein. J Biol Chem 260:7094–7101

Benovic JL, Kuhn H, Weyand I, Codina J, Caron MG & Lefkowitz RJ (1987a) Functional desensitization of the isolated β-adrenergic receptor by the β-adrenergic receptor kinase: potential role of the analog of the retinal binding protein arrestin (48KDa). Proc Natl Acad Sci USA 84:8879–8882

Benovic JL, Mayor F Jr, Staniszewski C, Lefkowitz RJ & Caron MG (1987b) Purification and characterization of the β-adrenergic receptor kinase. J Biol Chem 262:9026–9032

Bouvier M, Hausdorff WP, DeBlasi A, O'Dowd BF, Kobilka BK, Caron MG & Lefkowitz RJ (1988) Removal of phosphorylation sites from the β_2-adrenergic receptor delays onset of agonist-promoted desensitization. Nature (Lond) 333:370–372

Chung FZ, Wang CD, Porter PC, Venter JC & Fraser CM (1988) Site-directed mutagenesis and continuous expression of human β-adrenergic receptors. J Biol Chem 263:4052–4055

Clark RB (1986) Desensitization of hormonal stimuli coupled to regulation of cAMP levels. Adv Cyclic Nucleotide Protein Phosphorylation Res 20:151–209

Cotecchia S, Schwinn DA, Randall RR, Lefkowitz RJ, Caron MG & Kobilka BK (1988) Molecular cloning and expression of the cDNA for the hamster α_1-adrenergic receptor. Proc Natl Acad Sci USA 85:7159–7163

Dixon RAF, Sigal IF, Rands E, Register RB, Candelore MR, Blake AD & Strader CD (1987) Ligand binding to β-adrenergic receptor involves its rhodopsin-like core. Nature (Lond) 326:73–77

Dohlman HG, Caron MG, Strader CD, Amlaiky N & Lefkowitz RJ (1988) Identification and sequence of a binding site peptide of the β_2-adrenergic receptors. Biochem 27:1813–1817

Fargin A, Raymond J, Lohse MJ, Kobilka BK, Caron MG & Lefkowitz RJ (1988) The genomic clone G_{21} which resembles the β-adrenergic receptor sequence and encodes the 5HT1A receptor. Nature (Lond) 335:358–360

Hausdorff WP, Bouvier M, O'Dowd BF, Irons GP, Caron MG & Lefkowitz RJ (1989) Phosphorylation sites on two domains for the β_2-adrenergic receptor are involved in distinct pathways of receptor desensitization. J Biol Chem 264:12657–12665

Kobilka BK, Kobilka TS, Daniel K, Regan JW, Caron MG & Lefkowitz RJ (1988) Chimeric α_2-, β_2-adrenergic receptors: delineation of domains involved in effector coupling in ligand binding specificity. Science 240:1310–1316

Kunkel MW, Friedman J, Shenolikar S & Clark RB (1989) Cell-free heterologous desensitization of adenylyl cyclase in S_{49} lymphoma cell membranes mediated by cAMP dependent protein kinase. FASEB J 3:2067–2074

Lefkowitz RJ & Caron MG (1988) The adrenergic receptors: models for the study of receptors coupled to guanine nucleotide regulatory proteins. J Biol Chem 263:4993–4996

Liggett SB, Bouvier M, Hausdorff WP, O'Dowd BF, Caron MG & Lefkowitz RJ (1989) Altered patterns of agonist-stimulated cAMP accumulation in cells expressing mutant β_2-adrenergic receptors lacking phosphorylation sites. Mol Pharmacol. (in press)

Lohse MJ, Lefkowitz RJ, Caron MG & Benovic JL (1989) Inhibition of β-adrenergic receptor kinase prevents rapid homologous desensitization of the β_2-adrenergic receptors. Proc Natl Acad Sci USA 86:3011–3015

O'Dowd BF, Hnatowich M, Regan JW, Leader WM, Caron MG & Lefkowitz RJ (1988) Site directed mutagenesis of the cytoplasmic domains of the human β_2-adrenergic receptor. J Biol Chem 263:15985–15992

Regan JW, Kobilka TS, Yang Feng DL, Caron MG, Lefkowitz RJ & Kobilka BK (1988) Cloning and expression of a human kidney cDNA for a novel α_2-adrenergic receptor. Proc Natl Acad Sci USA 85:6301–6305

Rubinstein RC, Wang SKF & Ross EM (1987) The hydrophobic tryptic core of the β-adrenergic receptor retains G_s regulatory activity in response to agonists and thiols. J Biol Chem 262:16655–16662

Sibley DR, Benovic JL, Caron MG & Lefkowitz RJ (1987) Regulation of transmembrane signalling by receptor phosphorylation. Cell 48:913–922

Strader CD, Dixon RAF, Cheung AH, Candelore MR, Blake AD & Sigal IS (1987a) Mutation that uncoupled the β-adrenergic receptor from G_s and increased agonist affinity. J Biol Chem 262:16439–166443

Strader CD, Sigal IS, Register RV, Candelore MR, Rands E & Dixon RAF (1987b) Identification of residues required for ligand binding to the β-adrenergic receptors. Proc Natl Acad Sci USA 84:4384–4388

Strader CD, Sigal IS, Candelore MR, Rands E, Hill WS & Dixon RAF (1988) Conserved asparatic acid residues 79 and 113 of the β-adrenergic receptor have different roles in receptor function. J Biol Chem 63:4052–4055

Subject Index

MIX
Papier aus verantwortungsvollen Quellen
Paper from responsible sources
FSC® C105338

If you have any concerns about our products, you can contact us on
ProductSafety@springernature.com

In case Publisher is established outside the EU, the EU authorized representative is:
Springer Nature Customer Service Center GmbH
Europaplatz 3, 69115 Heidelberg, Germany

Printed by Libri Plureos GmbH
in Hamburg, Germany